MATRIX METHODS OF STRUCTURAL ANALYSIS

Element Approach

MATRIX METHODS OF STRUCTURAL ANALYSIS

Element Approach

S.S. Bhavikatti
Emeritus Professor
School of Civil Engineering
KLE Technological University
Hubballi (Karnataka)

Mamata B. Mogali
Assistant Professor
Department of Civil Engineering
SDM College of Engineering and Technology
Dharwad (Karnataka)

I.K. International Pvt. Ltd.
New Delhi

Published by
I.K. International Pvt. Ltd.
4435-36/7, Ansari Road, Daryaganj
New Delhi-110 002 (India)
E-mail: info@ikinternational.com
Website: www.techsarworld.com

ISBN: 978-93-90620-81-4

Published/Printed by Krishan Makhijani for I.K. International Pvt. Ltd., 4435-36/7, Ansari Road, Daryaganj, New Delhi-110 002.

Preface

Systematic development of consistent deformation method and slope deflection methods to suit the computer programming lead to flexibility matrix method and stiffness matrix method. Hence, in this book first consistent deformation and slope deflection methods are explained. Transformation of above methods to matrix method is best understood by element approach. Many universities insist on teaching element approach by solving simple problems by hand calculation. To meet this requirement authors have solved several problems.

Authors hope that the student community will wholeheartedly receive this book.

S.S. Bhavikatti
Mamata B. Mogali

Contents

1 Introduction

Determining the internal forces developed and deformations of the material of the members of a structure due to external forces acting on the structure is known as *structural analysis*. To design safe and economical structure, knowledge of structural analysis is required.

The various types of structures to be designed by a civil engineer may be classified into the following three groups:

1. One dimensional / skeleton
2. Two dimensional / plate
3. Three dimensional

One dimensional also known as skeleton or framed structures consist of one or more number of members which have one dimension (length) considerably larger compared to its cross-sectional dimensions. Examples of such structures are pin-jointed frames, rigid jointed frames, beams, and grids.

Two-dimensional structures, also known as plate structures, are slabs and walls which are subjected to loads at right angles to their surface. These structures have two dimensions considerably larger compared to their third dimension (thickness).

Three-dimensional structures have their dimensions in any direction considerably large. Examples of such structures are thick plates, soil mass under footing, etc.

The readers of this book are familiar with consistent deformation and slope deflection methods of structural analysis. The systematic development of these methods to suit computers application gave rise to *Matrix Method of Structural Analysis*. The development of consistent deformation method led to *Flexibility Matrix Method*, while the development of slope deflection method led to *Stiffness Matrix Method*.

In this chapter new technical terms used in matrix method of structural analysis are defined and explained.

1.1 STATIC INDETERMINACY

A structure which can be analysed with the use of equations of equilibrium of forces only is called *statically determinate structure*. Simply supported beams, beams with one end hinged and another on roller and cantilever beams are determinate beams. Similarly, pin-jointed plane frames with only three reaction components and number of members equal to $2j$-3 where j is the number of joints and single bay, single storey rigid frames with only three reaction components are determinate structures. Figure 1.1 shows some determinate structures.

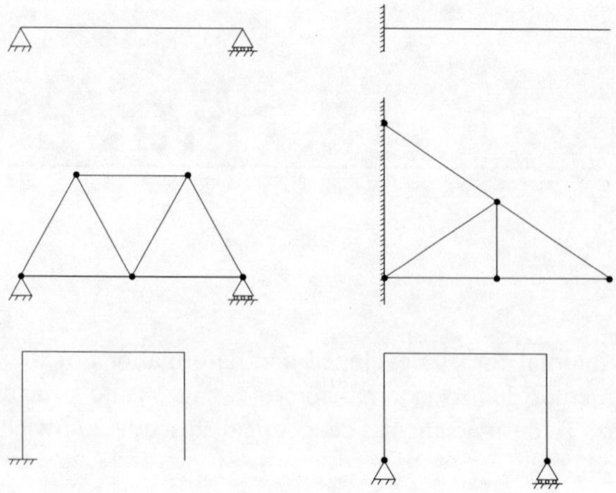

Fig. 1.1 Statically determinate structures.

A structure which cannot be analysed with equations of equilibrium only is a *statically indeterminate structure*. Figure 1.2 shows some indeterminate beams and frames, where support reactions are more than three. Figure 1.3 shows indeterminate frames where indeterminacy is due to more number of member forces than the number of available independent equations.

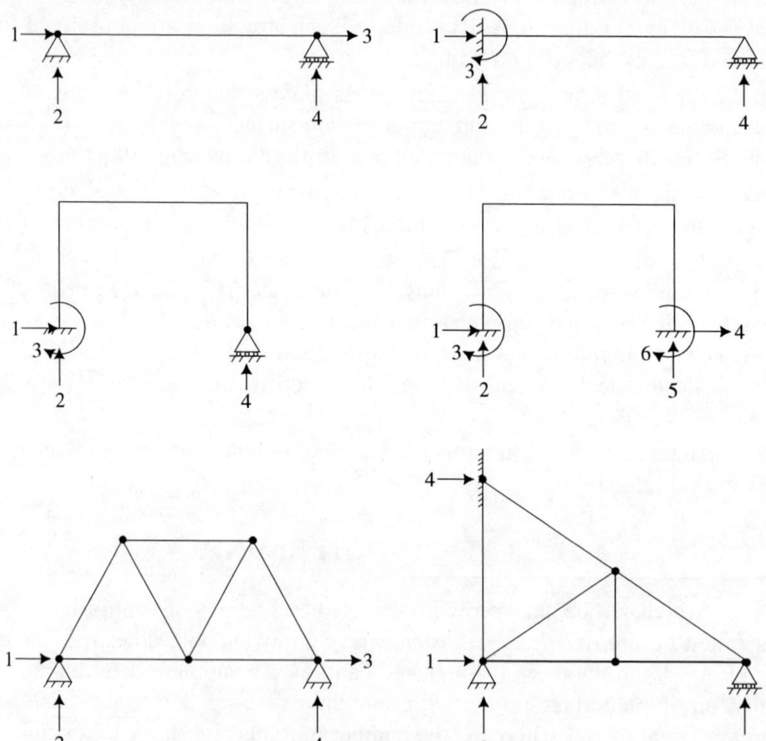

Fig. 1.2 Statically indeterminate structures.

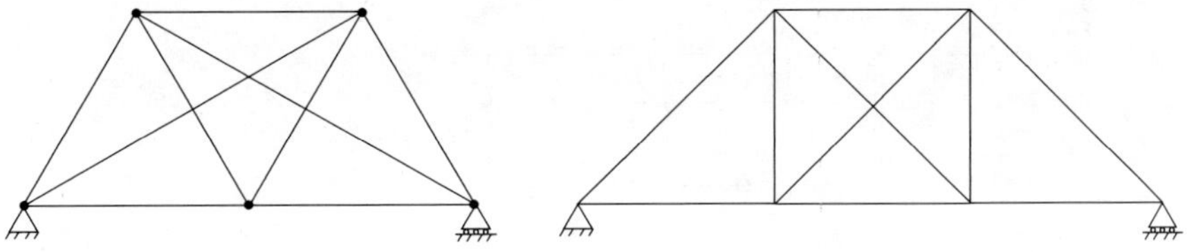

(a) (b)

Fig. 1.3 Indeterminate due to additional members.

1.2 DEGREE OF STATIC INDETERMINACY

The degree of static indeterminacy may be defined as additional equations to be formed using consistency of deformations so as to determine the redundant forces. Mathematically, it may be expressed as

Degree of static indeterminacy

= Number of redundant forces

= Number of unknown forces—Number of independent equations of equilibrium ...(1.1)

The method of determining the degree of static indeterminacy of the following structures is presented in this article:

1. Beams
2. Rigid jointed frames
3. Pin-jointed frames.

1.2.1 Beams

To find the number of unknown reactions, the following guidelines may be used:

Free end – 0

Roller end support – 1

Hinged end support – 2

Fixed end support – 3

Hence, total number of unknown reactions can be found.

There are only three independent equations of equilibrium for the beam analysis. However, if internal hinges are provided, there is one additional equation of equilibrium (moment at hinge = 0) for each hinge. Using these guidelines, the number of equilibrium equations can be determined. Then the degree of static indeterminacy may be found by using equation (1.1). Figure 1.4 shows the degree of static indeterminacy of the beams.

1.2.2 Rigid Jointed Frames

The method of determining the degree of static indeterminacy of plane frames and space frame is discussed below.

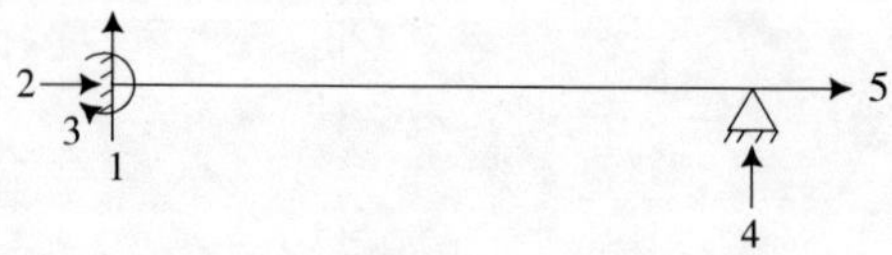

(a) $D_S = 5 - 3 = 2$

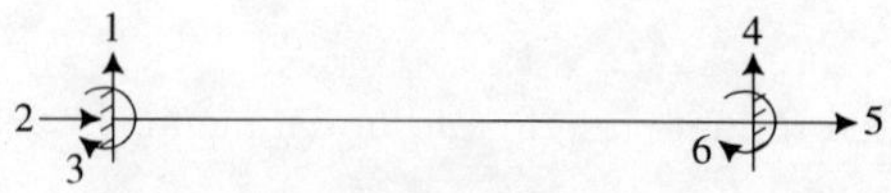

(b) $D_S = 6 - 3 = 3$

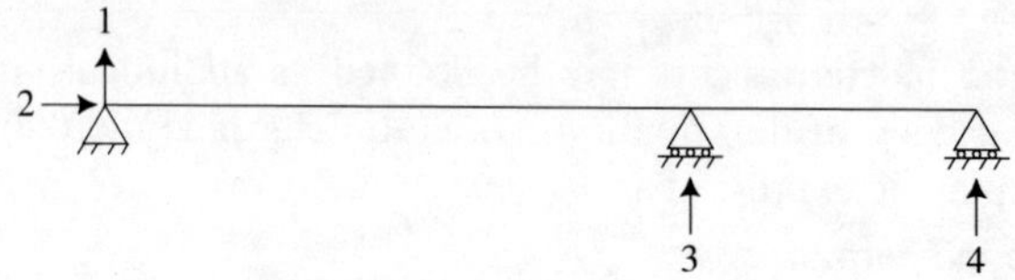

(c) $D_S = 4 - 3 = 1$

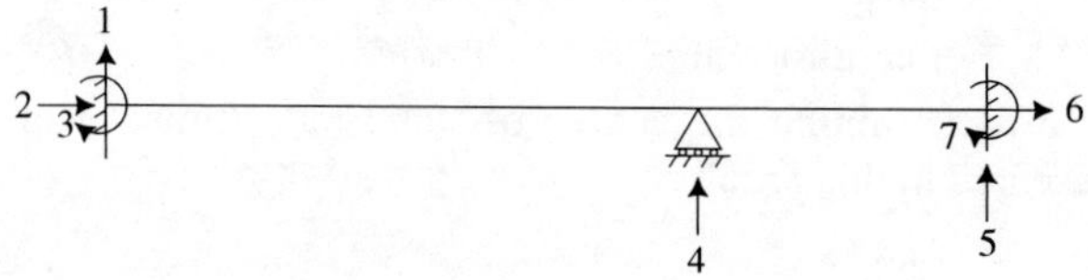

(d) $D_S = 7 - 3 = 4$

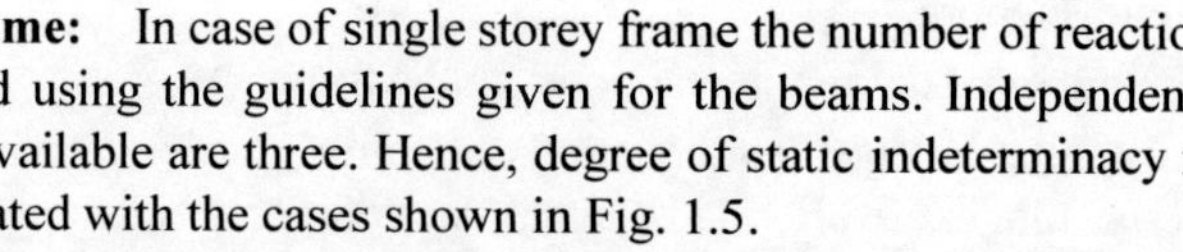

(e) $D_S = 7 - (3 + 2) = 2$

Fig. 1.4 Degree of static indeterminacy of beams.

(a) Plane Frame: In case of single storey frame the number of reaction components may be found using the guidelines given for the beams. Independent equations of equilibrium available are three. Hence, degree of static indeterminacy may be found. This is illustrated with the cases shown in Fig. 1.5.

However, in case of multi-storeyed frames, there are internal indeterminacies also. For example, consider the two-storeyed frame shown in Fig. 1.6. In this case even if all reaction components are determined, it is not possible to find bending moment and shear forces in the beams and columns of second storey. Hence, in this rigid jointed plane frame the degree of static indeterminacy is more than 3. To find the total degree of static indeterminacy, imagine a cut as shown in Fig. 1.7 (a) such that if the forces at each cut and reaction components are known, using equations of static equilibrium it is possible to find bending moment and shear force at any point in the frame. In this frame, number of unknown forces is equal to the number of reaction components plus three at each cut. The number of static equilibrium equations are three for each determinate portions. Knowing total number of unknown forces and static equilibrium equations, degree of static indeterminacy may be found. For the frame shown in Fig. 1.7(a),

Number of reaction components = 6

Number of cuts to make determinate portions = 2

$\therefore$ Number of unknowns at cuts = 3 × 2 = 6

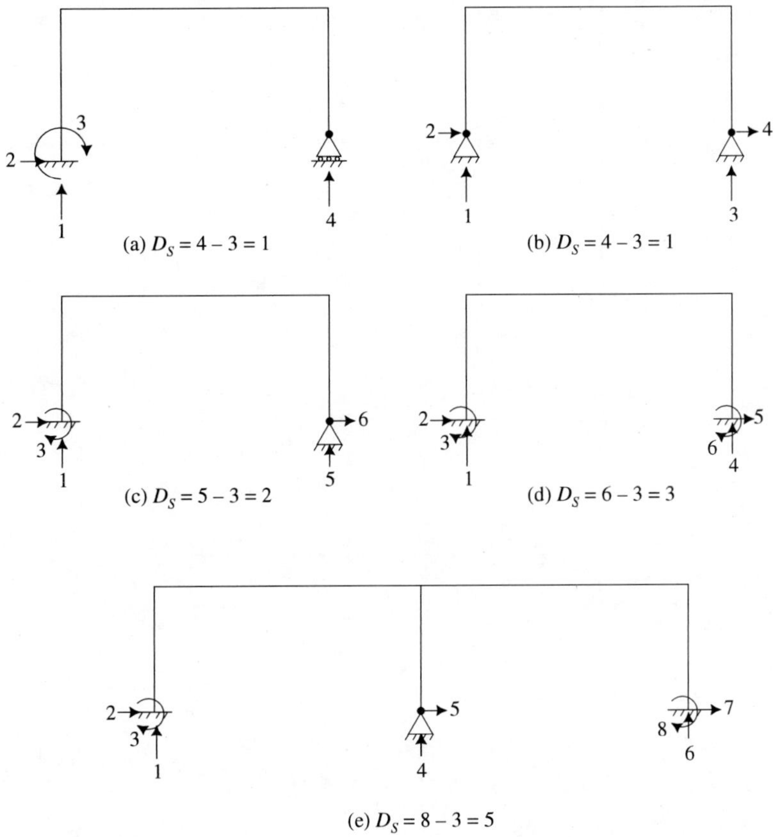

Fig. 1.5 Degree of static indeterminacy of single rigid jointed frames.

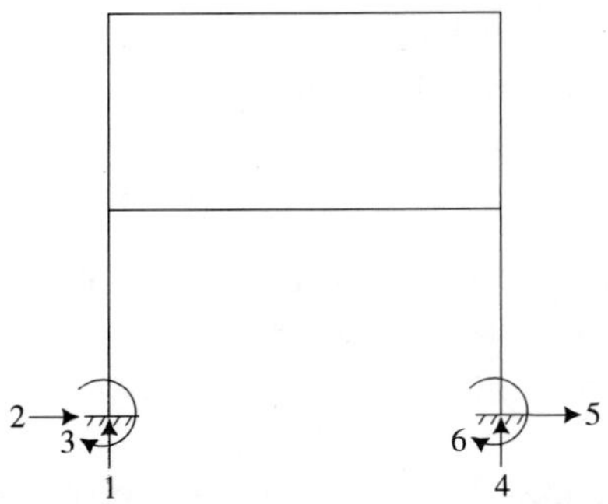

Fig. 1.6 A typical two-storey building.

∴ Total number of unknown forces = 6 + 6 = 12

Number of determinate portions = 2

∴ Number of independent equations of equilibrium = 3 × 2 = 6

Hence, degree of static indeterminacy, D_s = 12 – 6 = 6

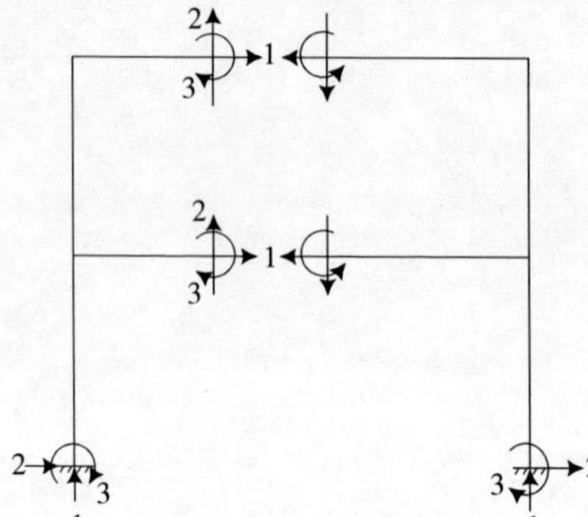

(a) $D_S = 3 \times 4 - 3 \times 2 = 6$

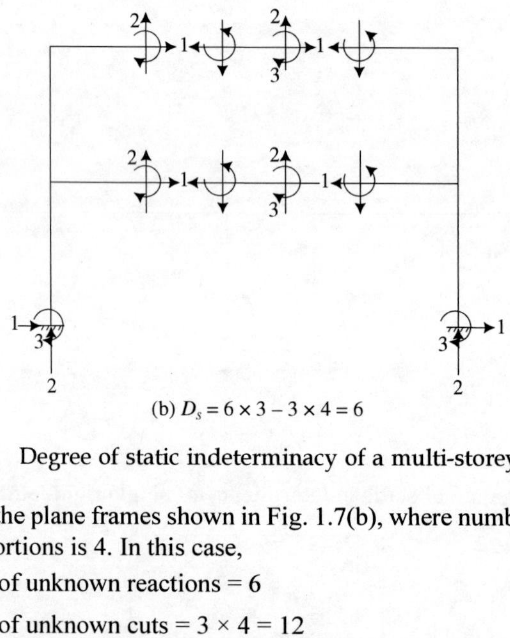

(b) $D_s = 6 \times 3 - 3 \times 4 = 6$

Fig. 1.7 Degree of static indeterminacy of a multi-storey frame.

Now consider the plane frames shown in Fig. 1.7(b), where number of cuts made to get determinate portions is 4. In this case,

Number of unknown reactions = 6

Number of unknown cuts = 3 × 4 = 12

∴ Number of unknown forces = 6 + 12 = 18

Number of independent static equilibrium equations = 3 × 4 = 12

∴ Degree of static indeterminacy = 18 –12 = 6

Thus, it is clear that any number of cuts may be made to get statically determinate portions.

Figure 1.8 illustrates the case of finding degree of static indeterminacy in a multistorey plane frame. In this,

Number of reaction components = 8

Number of cuts to get determinate portions = 6

Number of unknown at cuts = 3 × 6 = 18

Total number of unknown forces = 8 + 18 = 26

Number of determinate portions = 3

Number of independent equations of equilibrium = 3 × 3 = 9

Degree of static indeterminacy = 26 – 9 = 17

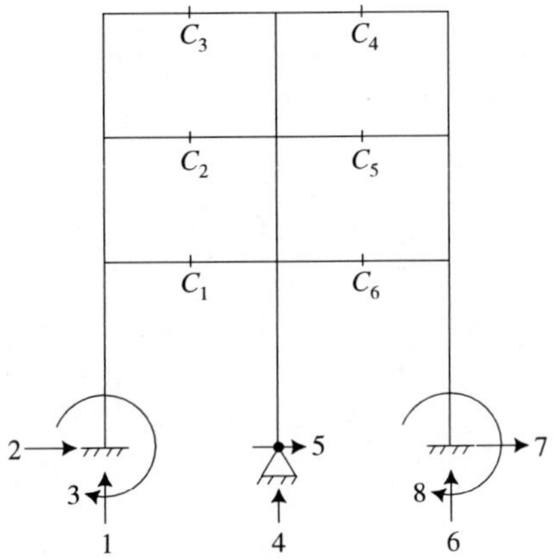

$D_s = (8 + 3 \times 6) - 3 \times 3 = 17$

Fig. 1.8 A three-storeyed frame.

(b) Space Frames: Number of reaction components at different supports are:

Roller end – 1(force normal to support)

Hinged end – 3 (force in *x*, *y* and *z* directions)

Fixed end – 6 (three force and three moment components)

However, if the roller movement is prevented in a particular direction in the plane of support, there will be one more reaction component in the direction movement is prevented. These are shown in Fig. 1.9.

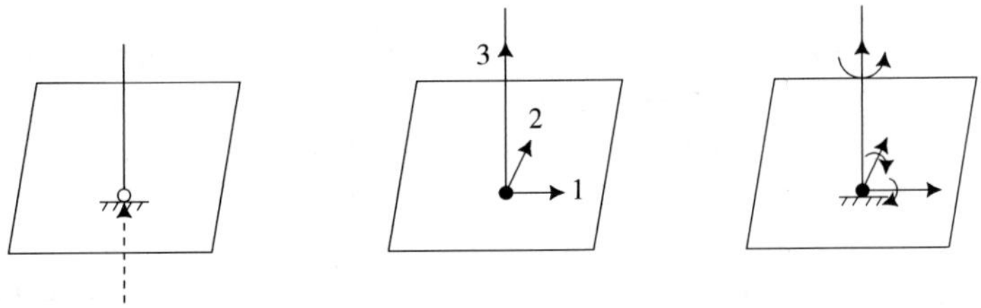

Fig. 1.9 Number of reaction components at support of a space frame.

To find the total degree of indeterminacy the technique of imagining sufficient cuts to get a number of determinate frames is used. In space frames, the number of unknown forces at each cut is six and the number of static equations of equilibrium for each determinate portion is six. Hence, the degree of static indeterminacy can be found. For example, consider the frame shown in Fig. 1.10. In this,

Number of reaction components = 6 × 6 = 36

∴ Number of cuts made to get determinate portions =14

Number of unknown cuts = 14 × 6 = 84

Number of determinate portions = 6

Number of independent equations = 6 × 6 = 36

Hence, degree of static indeterminacy, $D_s = (36 + 84) - 36 = 84$

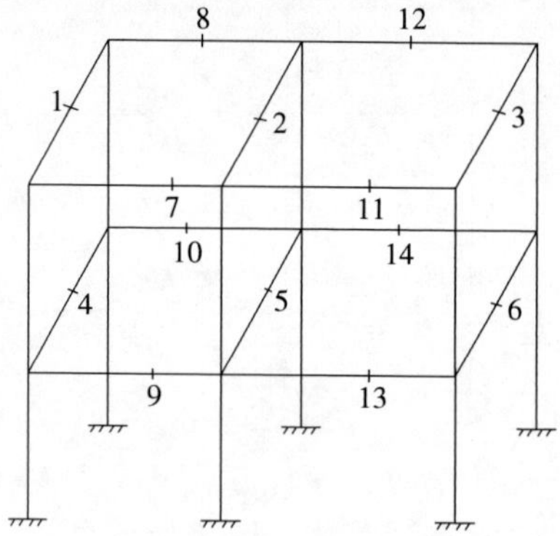

$$D_s = 6\times 6 + 6\times 14 - 6\times 6 = 84$$

Fig. 1.10 Degree of static indeterminacy of rigid jointed space frame.

1.2.3 Pin-jointed Frames

Method of finding degree of static indeterminacy of plane and space frames are discussed below.

(a) Plane Frames: Total degree of static indeterminacy may be divided into the following two groups:

(i) Degree of external indeterminacy (D_{se})

(ii) Degree of internal indeterminacy (D_{si})

(i) *Degree of external indeterminacy (D_{se}):* The difference between the number of reaction components and independent equilibrium gives the degree of external indeterminacy. For example, the degree of external indeterminacy in the frame shown in Fig. 1.11 is

$$D_{se} = 4 - 3 = 1.$$

(ii) *Degree of internal indeterminacy:* There are two independent equations of equilibrium for each joint. If the number of members is '*m*', there are '*m*' number of unknown forces. In externally determinate frames there are three reaction components.

Hence, total number of unknowns is $m + 3$. Therefore, the degree of internal indeterminacy is

$$D_{si} = m + 3 - 2j$$
$$= m - (2j - 3) \qquad \text{...(1.2)}$$

$\therefore$ Total degree of static indeterminacy

$$D_s = D_{se} + D_{si} \qquad \text{...(1.3)}$$

The degree of static indeterminacy of the frame shown in Fig. 1.11 may be found as given below:

$$D_{se} = 4 - 3 = 1$$

$$D_{si} = 11 - (2 \times 6 - 3) = 2$$

$\therefore$ $$D_s = 1 + 2 = 3$$

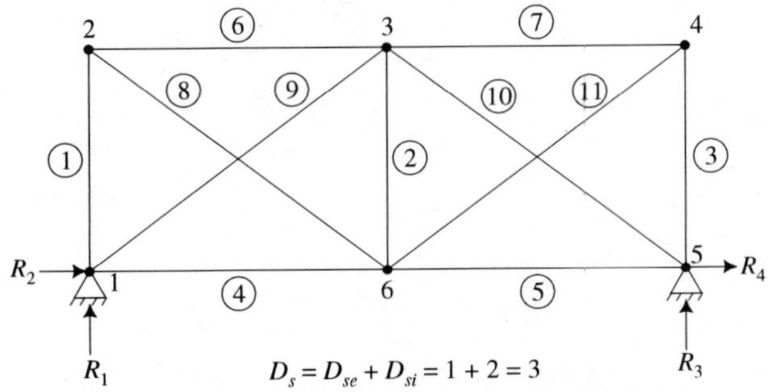

Fig. 1.11 Degree of static indeterminacy of a pin-jointed plane frame.

It may be noted that equation (1.2) is a necessary condition but not a sufficient condition. It is possible to satisfy the above condition by making a panel overstiff and a panel deficient. In general, a plane pin-jointed frame is a stable structure, if it consists of triangular panels without criss-crossing of members.

(b) Pin Connected Space Frames: In this case also we can divide total degree of indeterminacy into external and internal. It may be noted that in such frames external supports are either pin connected or on rollers. Total reaction components may be found as explained above. There are six independent equations of equilibrium for the entire structure. Hence, the degree of external indeterminacy is given by

$$D_{se} = r - 6 \qquad \text{...(1.4)}$$

where r is the number of reaction components.

There are three equations of equilibrium at each joint. Hence, if there are j number of joints, $3j$ number of equations can be formed. In an externally determinate frame there are six reaction components and force in each member is unknown. Hence, if 'm' is the number of members,

$$D_{si} = m + 6 - 3j$$
$$= m - (3j - 6) \qquad \text{...(1.5)}$$

Thus, in 25 members truss shown in Fig. 1.12 with four hinged supports

$$D_{se} = r - 6 = 3 \times 4 - 6 = 6$$

$$D_{si} = m - (3j - 6)$$

$$= 25 - (3 \times 10 - 6)$$

$$= 1.$$

∴ Total degree of static indeterminacy

$$D_s = D_{se} + D_{si}$$

$$= 6 + 1 = 7$$

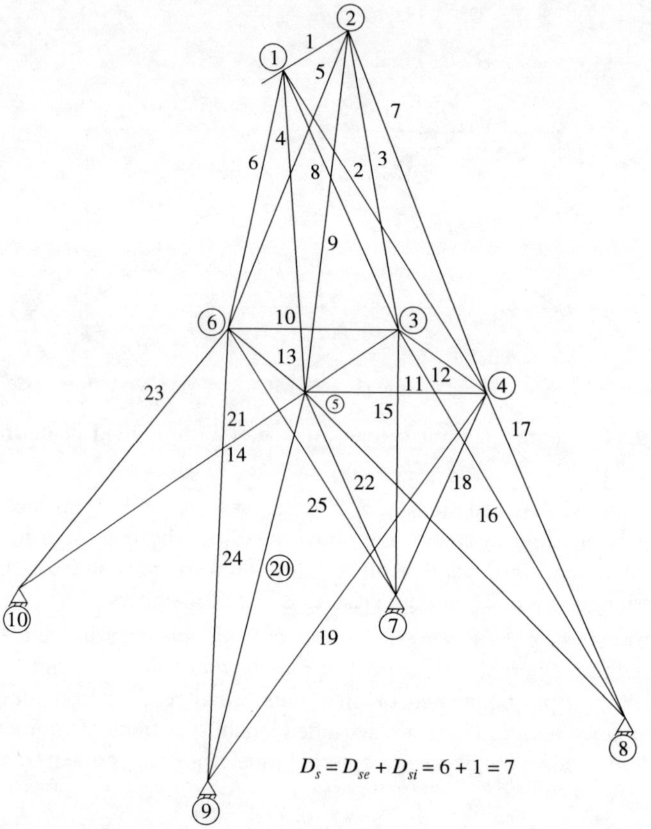

Fig. 1.12 Degree of static indeterminacy of a 25 bar truss.

1.3 KINEMATIC INDETERMINACY

If joint displacements of a member are known, the forces and moments in the member can be computed. This technique of structural analysis is used in slope deflection method. The unknown displacement components of a joint are known as degree of freedom of the joint. Total number of unknown displacement components of all joints in the structure is known as *degree of kinematic indeterminacy* of the structure. This is also known as *degree of freedom* of the structure. In a structure some of the displacement components are restrained to hold them in the position. Such restraints are known as *boundary conditions*. A skeleton structure is said to be *kinematically indeterminate* if all displacement components of its joints are not known by the boundary conditions alone. The methods of determining the degree of kinematic indeterminacy of the following skeleton structure is presented here:

1. Beams
2. Rigid frames
3. Pin-jointed frames

1.3.1 Degree of Kinematic Indeterminacy of Beams

At each end of a beam there are three displacement components, namely, one in axial direction, one at right angles to axis and a relation.

Hence, total degree of freedom of beam with '*j*' number of joints is $3j$. If '*e*', number of boundary conditions are known, the degree of indeterminacy of a beam is

$$D_k = 3j - e \qquad \text{...[1.6(a)]}$$

However, it has been found that axial deformations are negligible in beams. Hence, for beam with ends '*i*' and '*j*' axial deformation of joint '*i*' is the same as that of joint '*j*'. Thus, for each member of the beam one degree of freedom is reduced. Hence, for a beam with '*m*' number of members

$$D_k = 3j - e - m \qquad \text{...[1.6(b)]}$$

Table 1.1 shows the degree of freedom of some beams. It may be noted that for a fixed beam, $D_k = 3 \times 2 - 6 = 0$. In this case axial deformation is always zero.

Table 1.1 Degree of kinematic indeterminacy of beams

Beam	*Degree of kinematic indeterminacy*	
	Axial deformation	
	Considered	*Neglected*
	$3 \times 2 - 3 = 3$	$3 \times 2 - 3 - 1 = 2$
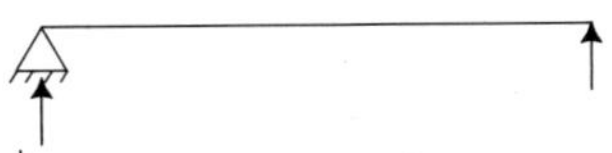	$3 \times 2 - 3 = 3$	$3 \times 2 - 3 - 1 = 2$
	$3 \times 2 - 4 = 2$	$3 \times 2 - 4 - 1 = 1$
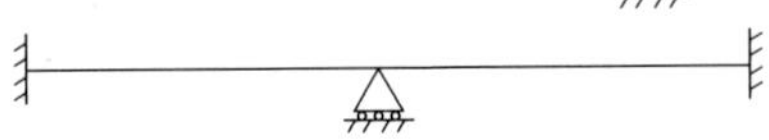	$3 \times 3 - 7 = 2$	$3 \times 3 - 7 - 1 = 1$
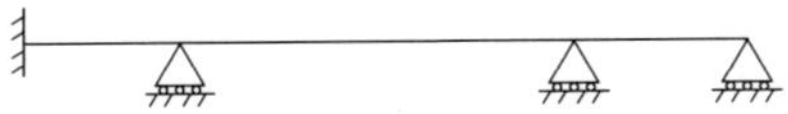	$3 \times 4 - 5 = 7$	$3 \times 4 - 5 - 3 = 4$
	$3 \times 4 - 5 = 7$	$3 \times 4 - 5 - 3 = 4$

1.3.2 Degree of Kinematic Indeterminacy of Rigid Jointed Frames

It may be easily seen that the degree of kinematic indeterminacy of rigid jointed plane frames is given by

$$D_k = 3j - e \quad \text{...[1.7(a)]}$$

For a member in space there are six displacement components at each joint, namely, 3 displacements in *x*, *y*, *z* coordinate directions and 3 rotations about coordinates. Hence, if '*e*' is the number of known boundary conditions, the degree of kinematic indeterminacy is given by

$$D_k = 6j - e \quad \text{...[1.7(b)]}$$

Since axial deformations are negligible in the rigid jointed frames, for each member one degree of indeterminacy reduces and hence

$$D_k = 3j - e - m \text{ for plane frames} \quad \text{...[1.8(a)]}$$

$$= 6j - e - m \text{ for space frames} \quad \text{...[1.8(b)}$$

Figure 1.13 shows the degree of kinematic indeterminacy of some of the rigid jointed plane frames and Fig. 1.14 shows that of a rigid space frame treating axial deformation is negligible.

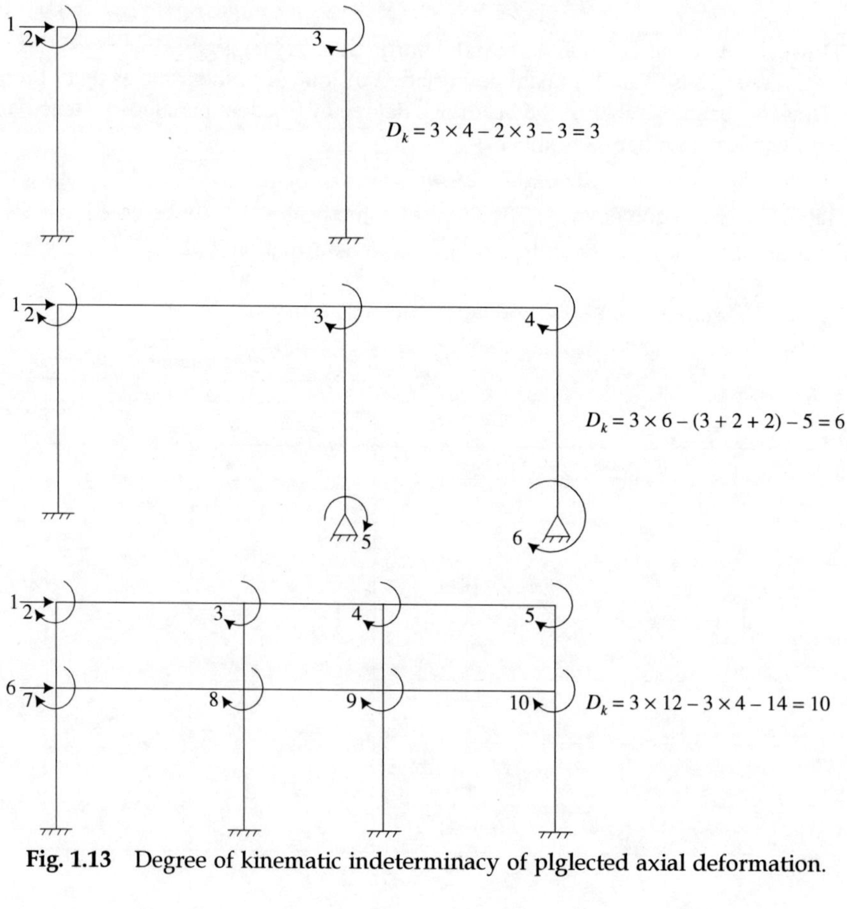

Fig. 1.13 Degree of kinematic indeterminacy of plglected axial deformation.

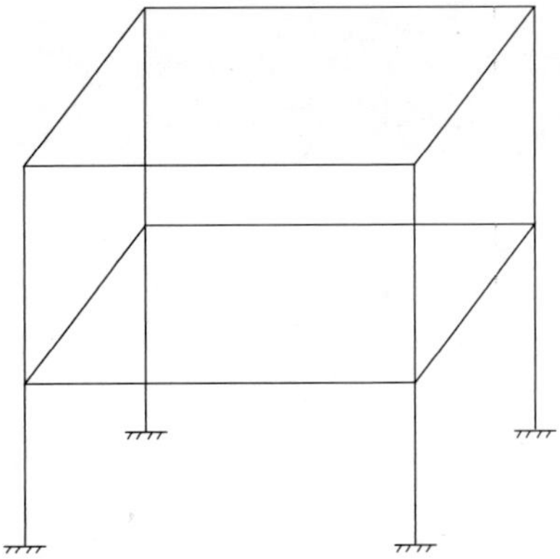

$D_k = 6 \times 12 - 6 \times 4 - 16 = 32$

Fig. 1.14 Degree of freedom of rigid jointed space frame neglecting axial deformation.

1.3.3 Degree of Kinematic Indeterminacy of Pin-jointed Frames

In case of pin-jointed frames connections permit free rotation of joints. Hence, rotations do not cause any internal force. Displacements of joints cause member forces. Hence, displacements in Cartesian coordinates are of interest in the analysis. Hence, in a plane frame the degree of kinematic indeterminacy (degree of freedom) of a joint is two in case of plane frames and three in case of space frames. If e is the number of known boundary conditions, then the degree of kinematic indeterminacy is given by

$$D_k = 2j - e \text{ for pin-jointed plane frames} \quad ...[1.9(a)]$$

and

$$D_k = 3j - e \text{ for pin-jointed space frames} \quad ...[1.9(b)]$$

where j is number of joints.

Figure 1.15 shows the degree of kinematic indeterminacy of some of the pin connected frames.

1.4 GENERALIZED COORDINATES

The direction of forces/deformations which are to be determined to know the member forces are known as *generalized coordinates.* These coordinates may be numbered arbitrarily. However, systematic numbering helps in developing general-purpose analysis programs to a great extent. For hand calculations the direction of only unknowns may be selected as coordinates. But in case of developing computer programs, it is convenient to take the direction of all forces/all displacements of joints as coordinate directions. Figure 1.16 shows some of the methods of selecting generalized coordinates. However, usually preferred system is the one shown in Fig. 1.16(b).

1.5 GENERALIZED FORCE AND DEFORMATION

In matrix analysis force and deformations are generalized terms. Force means forces as well as moment. Similarly, the term deformation is used to mean deflection and rotation. Thus in Fig. 1.16(b):

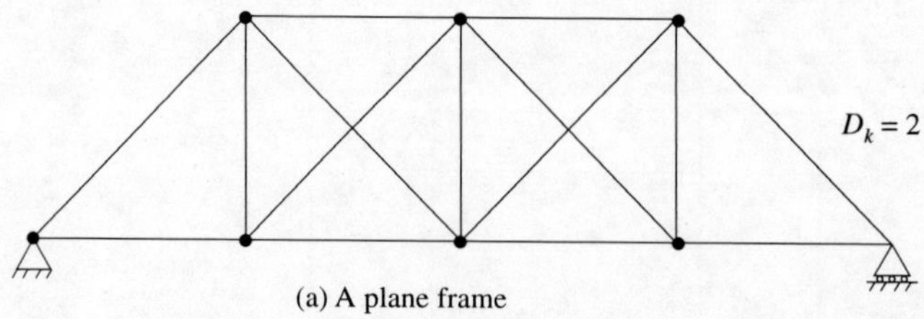

$$D_k = 2 \times 8 - 3 = 13$$

(a) A plane frame

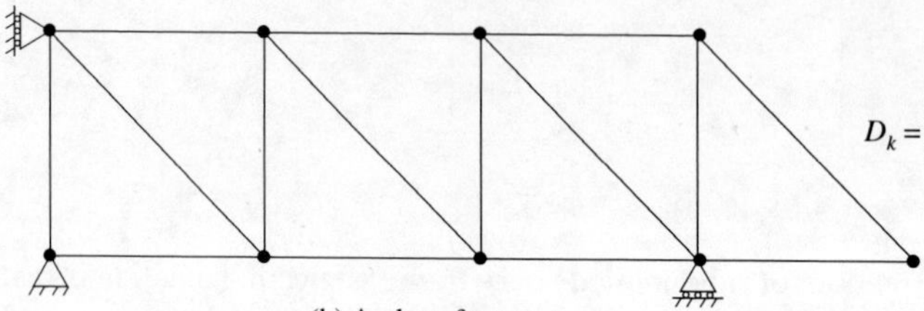

$$D_k = 2 \times 9 - 4 = 14$$

(b) A plane frame

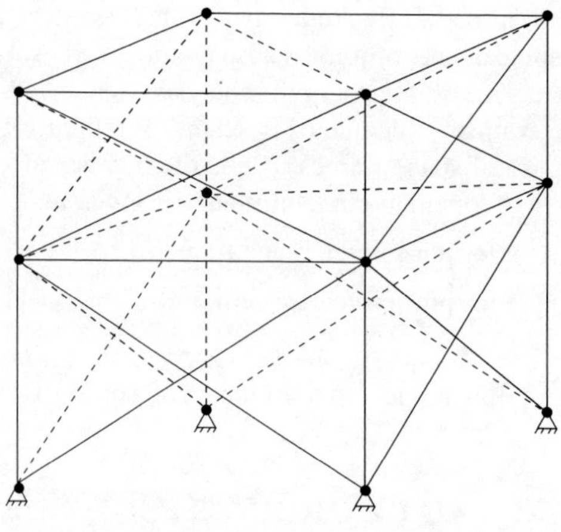

$$D_k = 3 \times 12 - 2 \times 4 = 28$$

(c) 32-Member space frame

Fig. 1.15 Degree of kinematic indeterminacy of pin-jointed frames.

F_4 is horizontal force at joint B

Δ_4 is horizontal displacement at B

F_9 is the moment at joints C and

Δ_9 is the rotation at C.

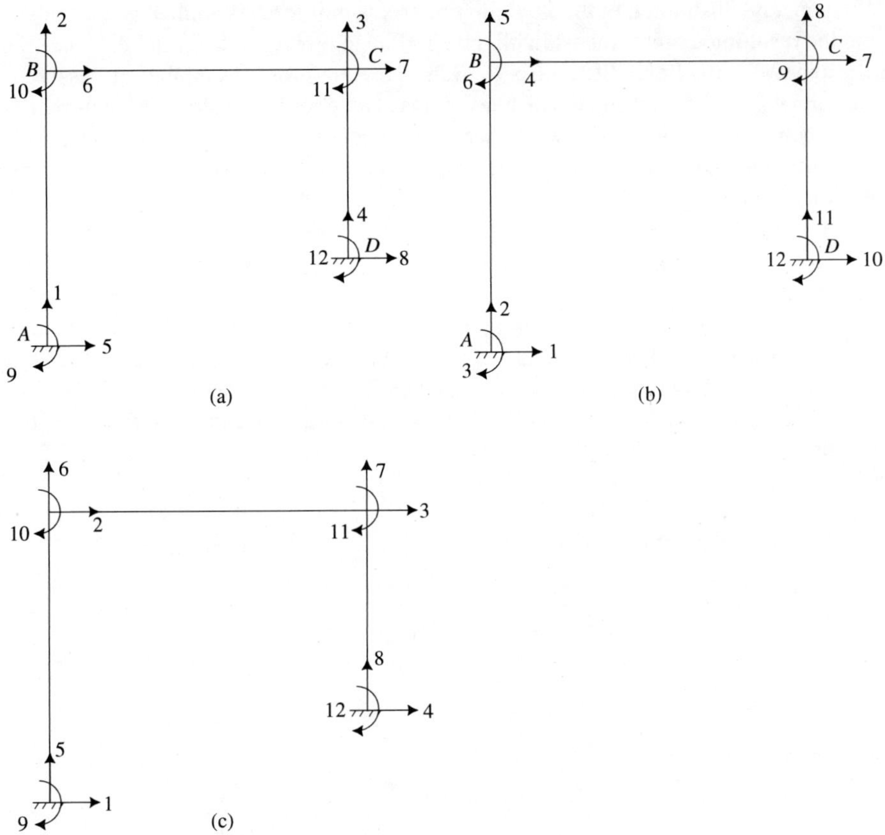

Fig. 1.16

1.6 FLEXIBILITY MATRIX

If a structure has n number of coordinates, its displacement response to the forces is represented by

$$[\delta] = \begin{bmatrix} \delta_{11} & \delta_{12} & \delta_{13} & \cdots & \delta_{14} \\ \delta_{21} & \delta_{22} & \delta_{23} & \cdots & \delta_{24} \\ \vdots & \vdots & \vdots & & \vdots \\ \delta_{n1} & \delta_{n2} & \delta_{n3} & \cdots & \delta_{n4} \end{bmatrix} \qquad ...(1.10)$$

The matrix $[\delta]$ is known as *flexibility matrix*. The element δ_{ij} is the displacement at coordinate i due to a unit force at coordinate j.

To generate *j*th column of the flexibility matrix, a unit force is applied at coordinate *j* and the resulting displacements in all the coordinate direction determined. Hence, in order to develop the flexibility matrix for a structure, unit force is applied successively at coordinates 1, 2, 3 … *n* and the displacements at all coordinate directions computed. It is to be noted that from Maxwell's reciprocal theorem, $\delta_{ij} = \delta_{ji}$. Hence, flexibility matrix has diagonal symmetry and this property is used commonly while developing flexibility matrix.

1.7 FLEXIBILITY EQUATION

Consider a structure subjected to forces $\{F\}^T = [F_1\ F_2 \ldots F_n]$ where, F_i is the force in direction *i*. Let it have displacements $\{\Delta\}^T = [\Delta_1\ \Delta_2 \ldots \Delta_n]$. Since δ_{ij} is the displacement in coordinate direction *i* due to unit force in coordinate *j*, the displacement at *i* due to force F_1 is $\delta_{11} F_1$. Similarly, $\delta_{12} F_2$ is the displacement at *i* due to force F_2. Thus, the total displacement is given by

$$\delta_{i1}F_1 + \delta_{i2}F_2 + \delta_{i3} F_2 + \cdots + \delta_{in}F_n = \Delta_i$$

If the above expression is written for all displacement, we have

$$\begin{array}{llllll}
\delta_{11} F_1 & + \delta_{12}F_2 & + \delta_{13}F_3 & + \cdots & + \delta_{1n}F_n & = \Delta_1 \\
\delta_{21}F_1 & + \delta_{22}F_2 & + \delta_{23}F_3 & + \cdots & + \delta_{2n}F_n & = \Delta_2 \\
\vdots & \vdots & \vdots & & \vdots & \vdots \\
\delta n_1 F_1 & + \delta n_2 F_2 & + \delta n_3 F_3 & + \ldots & + \delta_{nn}F_n & = \Delta_n
\end{array}$$

i.e., in matrix form

$$\begin{bmatrix} \delta_{11} & \delta_{12} & \delta_{13} & \cdots & \delta_{1n} \\ \delta_{21} & \delta_{22} & \delta_{23} & \cdots & \delta_{2n} \\ \vdots & & & & \\ \delta_{n1} & \delta_{n2} & \delta_{n3} & \cdots & \delta_{nn} \end{bmatrix} \begin{Bmatrix} F_1 \\ F_2 \\ \vdots \\ F_n \end{Bmatrix} = \begin{Bmatrix} \Delta_1 \\ \Delta_2 \\ \vdots \\ \Delta_n \end{Bmatrix}$$

i.e. $$[\delta]\ \{F\} = \{\Delta\} \qquad \ldots(1.11)$$

1.8 STIFFNESS MATRIX

If a structure has *n* number of coordinates, its force response to the displacement is represented by

$$[k] = \begin{bmatrix} k_{11} & k_{12} & k_{13} & \cdots & k_{1n} \\ k_{21} & k_{22} & k_{23} & \cdots & k_{2n} \\ \vdots & \vdots & \vdots & & \vdots \\ k_{41} & k_{42} & k_{43} & \cdots & k_{nn} \end{bmatrix}$$

and the matrix [*k*] is known as stiffness matrix. The element k_{ij} is the force developed at coordinate *i* due to a unit displacement at coordinate *j*. Hence, to develop *j*th column of the stiffness matrix a unit displacement is given at *j*th coordinate and the forces developed at all coordinates are determined. Thus, to develop the stiffness matrix unit displacements should be given successively at coordinates 1, 2, … *n*, and the forces developed at all coordinates computed. Noting that according to Maxwell's reciprocal theorem k_{ij} is equal to k_{ji}, the stiffness matrix is a symmetric matrix about its diagonal. This property is also used while developing stiffness matrix.

1.9 STIFFNESS EQUATION

Let F_i be the force and Δ_i be the displacement at coordinate i since k_{ij} is the force at joint i due to unit displacement at j, $k_{i1}\Delta_1$ is the force in coordinate i due to displacement Δ_1. Similarly, $k_{i2}\Delta_2$ is the force in coordinate j due to displacement Δ_2. Thus, the total force F_i is given by

$$k_{i1}\Delta_1 + k_{i2}\,\Delta_2 + k_{i3}\Delta_3 + \dots + k_{in}\Delta_n = F_i$$

Thus, we have

$$\begin{array}{l} k_{11}\Delta\ + k_{12}\Delta_2\ + k_{13}\Delta_3 + \dots + k_{1n}\Delta_n = F_1 \\ k_{21}\Delta_1 + k_{22}\Delta_2\ + k_{23}\Delta_3 + \dots + k_{2n}\Delta_n = F_2 \\ \vdots \quad\quad \vdots \quad\quad \vdots \quad\quad\quad\quad \vdots \quad\quad \vdots \\ k_{n1}\Delta_1 + k_{n2}\Delta_2\ + k_{n3}\Delta_3 + \dots + k_{nn}\Delta_n = F_n \end{array}$$

i.e.

$$[k]\,\{\Delta\} = \{F\} \quad \text{...(1.12)}$$

1.10 RELATIONSHIP BETWEEN FLEXIBILITY AND STIFFNESS MATRICES

From equation 1.12,

$$[k]\,\{\Delta\} = \{F\}.$$

Pre-multiplying both sides with $[k]^{-1}$, we get

$$\{\Delta\} = [k]^{-1}\{F\}$$

But from equation 1.11, we have

$$[\delta]\,\{F\} = \{\Delta\}.$$

Hence,

$$[\delta]\,\{F\} = [k]^{-1}\{F\}.$$

or

$$[\delta] = [k]^{-1} \quad \text{...(1.13)}$$

Thus, *flexibility and stiffness matrices are inverse of each other.*

Questions

1. Briefly explain the terms static and kinematic indeterminacy of structures.
2. Distinguish between
   (a) Static and kinematic indeterminacy.
   (b) External and internal indeterminacy.
3. Determine the degree of static and kinematic indeterminacy of the continuous beams shown in Fig. 1.17. Treat members as inextensible [axial deformations neglected].

(a) Problem 1.3 (a) **Ans:** $D_s = 2$
$D_k = 4$

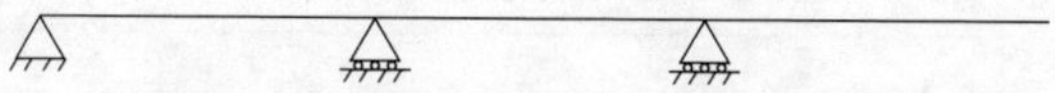

(b) Problem 1.3 (b) **Ans:** $D_s = 1$
$D_k = 5$

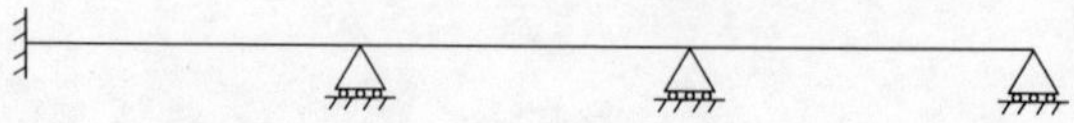

(c) Problem 1.3 (c) **Ans:** $D_s = 3$
$D_k = 3$

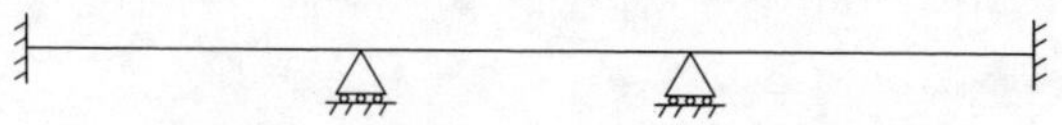

(d) Problem 1.3 (d) **Ans:** $D_s = 5$
$D_k = 2$

Fig. 1.17

4. Determine the degree of static and kinematic indeterminacy of the plane frames shown in Fig. 1.18. Neglect axial deformations in rigid frames.

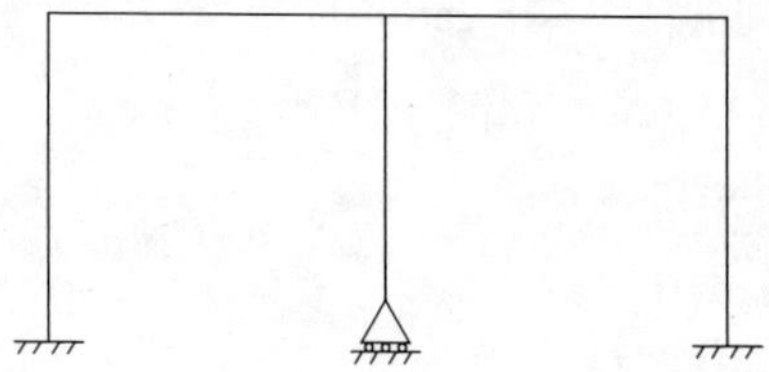

(a) Problem 1.4 (a) **Ans:** $D_s = 4$
$D_k = 6$

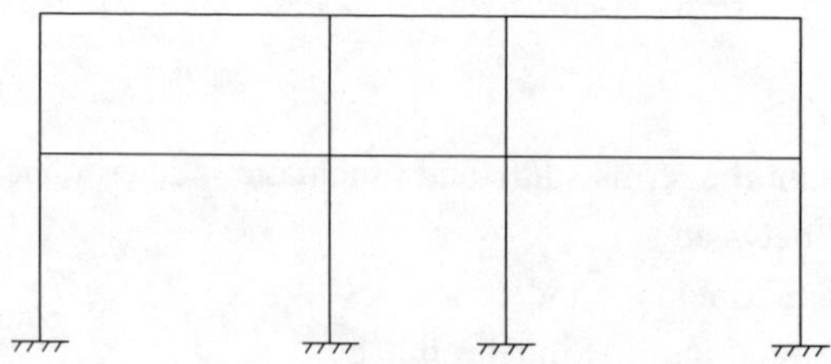

(b) Problem 1.4 (b) **Ans:** $D_s = 18$
$D_k = 10$

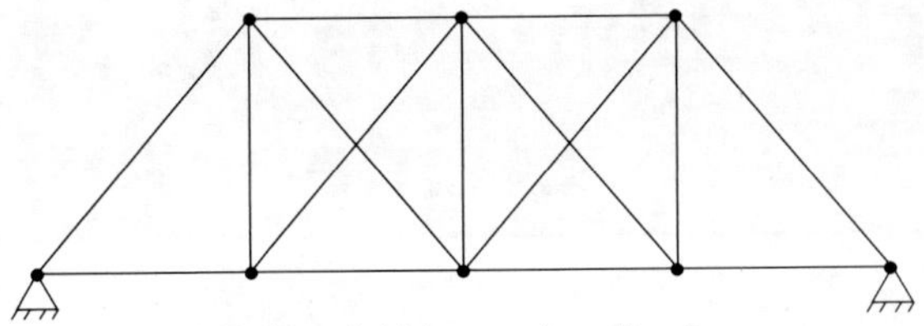

(c) Problem 1.4 (c) **Ans:** $D_s = 3$
$D_k = 12$

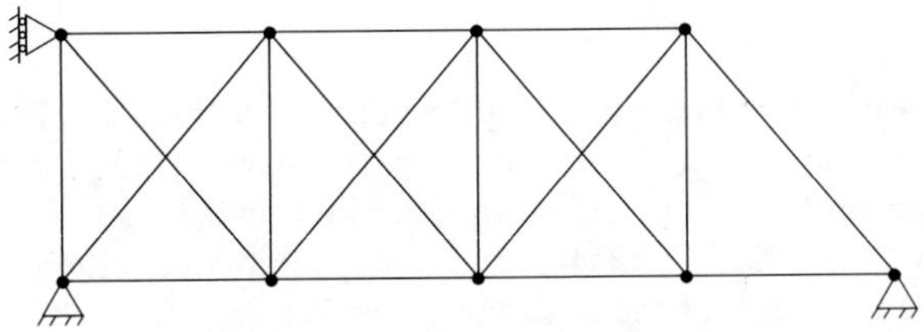

(d) Problem 1.4 (d) **Ans:** $D_s = 4$
$D_k = 14$

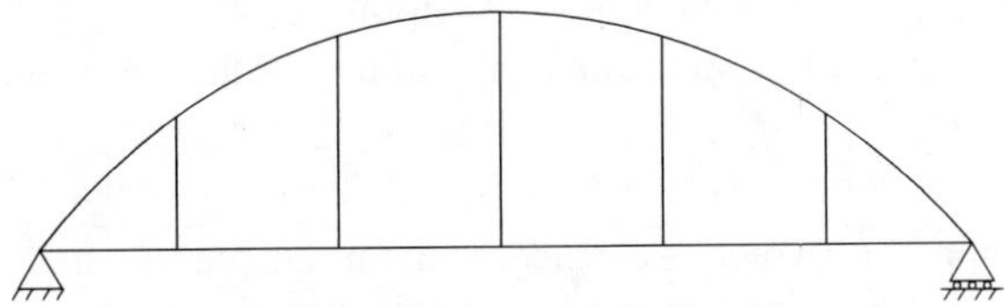

(e) Problem 1.4 (e) **Ans:** $D_s = 18$
$D_k = 16$

Fig. 1.18

5. Determine the degree of indeterminacy of the space frame shown in Fig. 1.19. Neglect axial deformations.

Ans: $D_s = 48$
$D_k = 48$

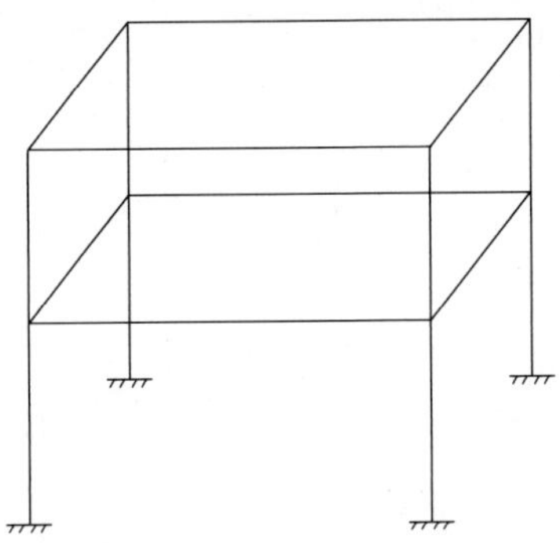

(a) Problem 1.5 (a)

Ans: $D_s = 8$
$D_k = 24$

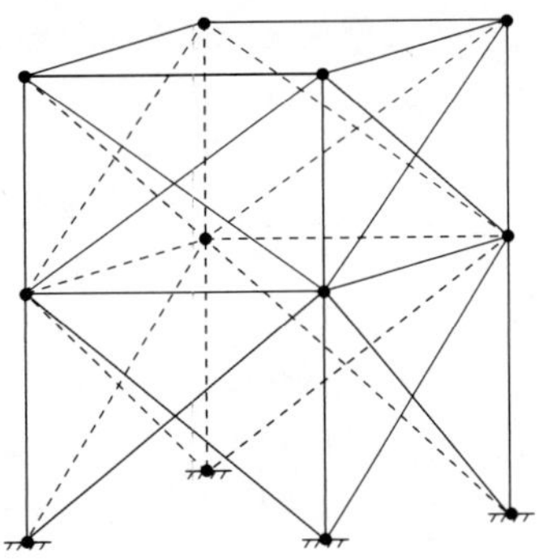

(b) 32-Member frame

Fig. 1.19

2 Flexibility Matrix Method: Element Approach

The behaviour of the structure can largely be defined by force-displacement relation in the form of a matrix. The systematic development of consistent deformation method in the matrix form has resulted in flexibility matrix method. In the method, the basic unknowns are the redundant forces, compatibility conditions are enforced at the discrete points throughout the structure to get the set of simultaneous equations in which the coefficients are flexibility coefficients. These equations written in the matrix form are solved for the unknown forces. This method is known as:

1. Force method since redundant forces are to be taken as unknown.
2. Compatibility method since the compatibility or the consistency equations are enforced.
3. Flexibility method since flexibility coefficients are used.

The element approach which is considered in this chapter is suitable for developing general-purpose computer program. In this chapter flexibility matrix method by element approach is explained and is used to solve the following types of one-dimensional elements with which readers are familiar to solve:

(a) Beams
(b) Rigid jointed plane frames
(c) Pin-jointed plane frames

2.1 FLEXIBILITY EQUATION

Flexibility is defined as displacement due to unit force.

Flexibility Matrix: Equation used: $[\delta]\{F\} = \{\Delta\}$

where,

$\{F\}$ = Force vector

$$\{F\} = \begin{Bmatrix} F_1 \\ F_2 \\ F_3 \\ \cdot \\ \cdot \\ F_n \end{Bmatrix}$$

These are generalized forces. They may be axial, translation, moment or twist.

$\{\Delta\}$ = Displacement vector

$$\{\Delta\} = \begin{Bmatrix} \Delta_1 \\ \Delta_2 \\ \Delta_3 \\ . \\ . \\ \Delta_n \end{Bmatrix}$$

These are generalized displacements. They may be axial, translation, moment or twist.
$[\delta]$ = Flexibility matrix

$$[\delta] = \begin{bmatrix} \delta_{11} & \delta_{12} & . & . & \delta_{1n} \\ \delta_{21} & \delta_{22} & . & . & \delta_{2n} \\ . & . & \delta_{ij} & . & . \\ \delta_{n1} & \delta_{n2} & . & . & \delta_{nn} \end{bmatrix}$$

where, n = no. of coordinates

δ_{ij} = flexibility coefficient

Flexibility coefficient is defined as the displacement at coordinate i due to unit force at coordinate j, i.e., δ_{ij} .

Example 2.1 Develop the flexibility matrix for the beam shown in Fig. 2.1 having constant EI.

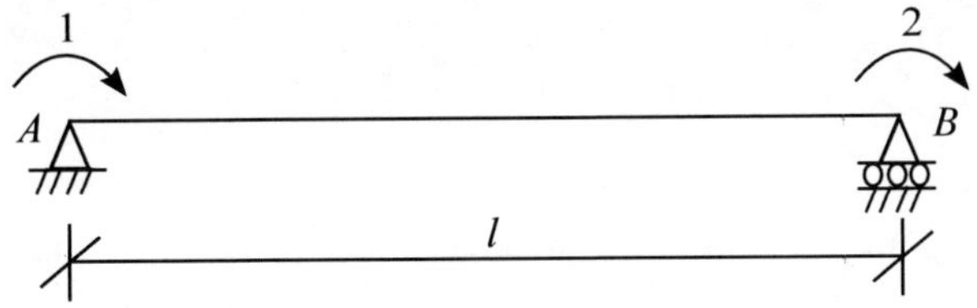

Fig. 2.1

Solution:

$$[\delta] = \begin{bmatrix} \delta_{11} & \delta_{12} \\ \delta_{21} & \delta_{22} \end{bmatrix}$$

To get the first column of the matrix, apply unit force along coordinate 1 and find the displacements at 1 and 2.

Apply unit force along coordinate 1 as shown in Fig. 2.2.

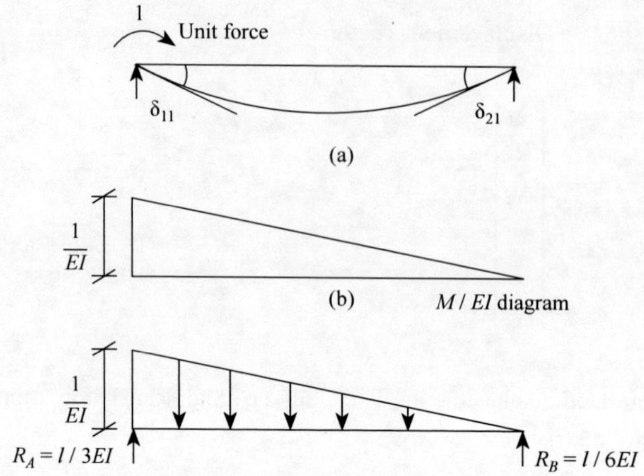

(c) Conjugate beam loaded with M/EI diagram

Fig. 2.2

Slope at any section of the given beam = Shear force of the conjugate beam

$$\Sigma M_B = 0$$

$$R_A \times l - \left(\frac{1}{2} \times l \times \frac{1}{EI} \times \frac{2}{3} \times l\right) = 0$$

$$R_A = \frac{l}{3EI}$$

$$\Sigma V = 0$$

$$R_A + R_B = \frac{1}{2} \times l \times \frac{1}{EI}$$

$$R_B = \frac{l}{6EI}$$

δ_{11} = Slope at the left support of the given beam = Shear force at the left support of the conjugate beam = $\frac{l}{3EI}$

δ_{21} = Slope at the right support of the given beam = Shear force at the right support of the conjugate beam = $-\frac{l}{6EI}$

Similarly,

To get the second column of the matrix, apply unit force along coordinate 2 and find the displacements at 1 and 2.

Apply unit force along coordinate 2 as shown in Fig. 2.3.

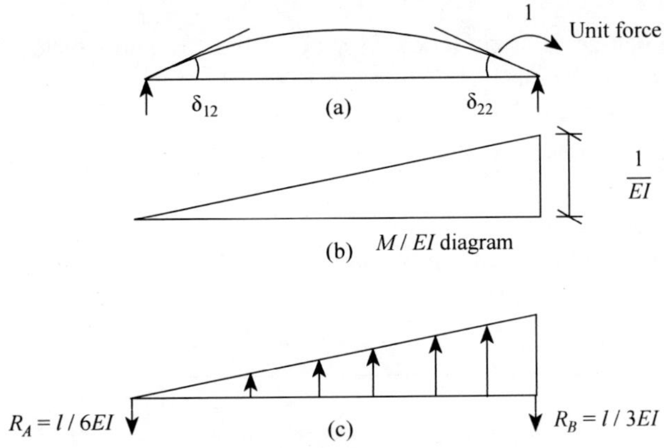

Conjugate beam loaded with M/EI diagram

Fig. 2.3

$$\delta_{12} = -\frac{l}{6EI}$$

$$\delta_{22} = \frac{l}{3EI}$$

$$[\delta] = \begin{bmatrix} \frac{l}{3EI} & -\frac{l}{6EI} \\ -\frac{l}{6EI} & \frac{l}{3EI} \end{bmatrix}$$

$$[\delta] = \frac{l}{6EI}\begin{bmatrix} 2 & -1 \\ -1 & 2 \end{bmatrix}$$

2.2 ANALYSIS OF BEAMS

Continuous beams are analysed here by flexibility matrix method.

Example 2.2 Analyze the continuous beam shown in Fig. 2.4(a) by flexibility matrix method.

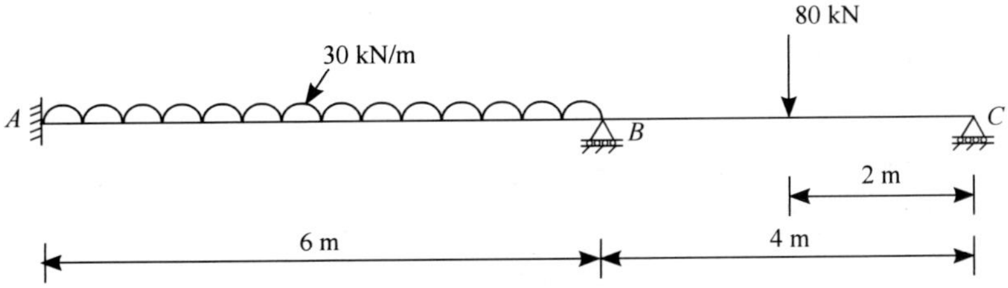

Fig. 2.4(a)

Solution:

Step 1:

Determine the degree of static indeterminacy of the structural system. Identify the redundants and the primary structure or released structure. Assign coordinates 1, 2, ..., *n* to the chosen redundants.

In this beam, the degree of static indeterminacy = 2. Here the axial deformations are neglected.

The primary structure is chosen as shown below, by introducing hinges at *A* and *B*. The moments at *A* and *B* are thus redundants. The moments at *A* and *B* are redundant forces as shown in Fig. 2.4(b).

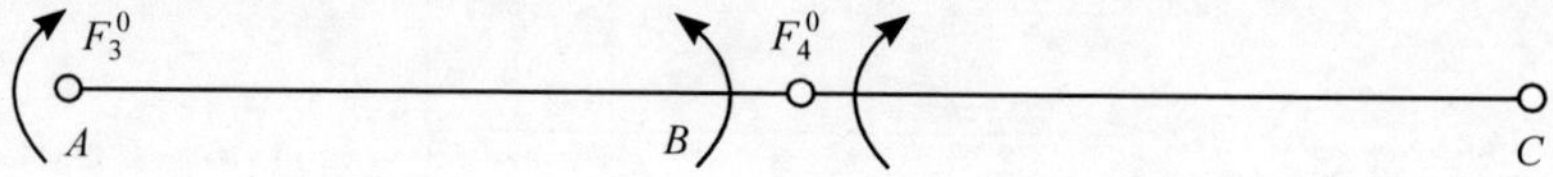

Primary structure with redundant forces

Fig. 2.4(b)

$$\{F^0\} = \begin{Bmatrix} F_3^0 \\ F_4^0 \end{Bmatrix}$$

where, $\{F^0\}$ = redundant forces

Step 2:

Identify the elements of the structural system and assign the element coordinates.

We have to divide the structure into number of elements and fix P-coordinates. Mark all the coordinates in clockwise direction.

In this problem, the beam has been divided into two elements and we have chosen four element coordinates as shown in Fig. 2.4(c).

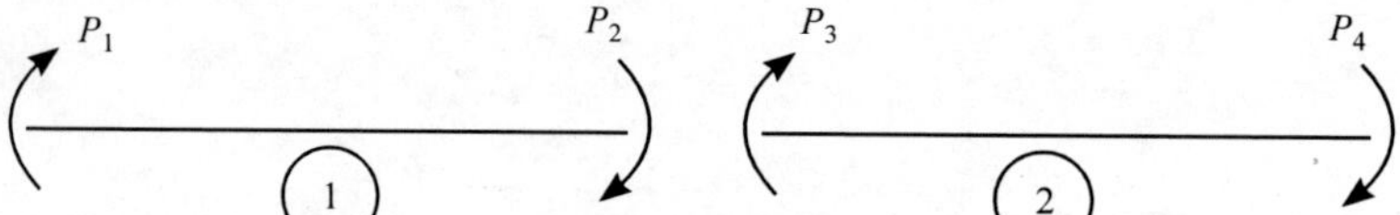

Element coordinates

Fig. 2.4(c)

$$\{P\} = \begin{Bmatrix} P_1 \\ P_2 \\ P_3 \\ P_4 \end{Bmatrix}$$

where, $\{P\}$ = Element force vector.

Step 3:

Compile the force vector, $\{F\}$ which comprises $\{F^*\}$ and $\{F^0\}$.

Compile the external forces, $\{F^*\}$ where, $\{F^*\}$ is called system coordinate.

System coordinates are the forces resulting from the given external forces. Now the system forces must be transferred to coordinates. We have to find equivalent joint forces. Hence, if any load acting at a point other than the joint, then it should be replaced by its equivalent load at the joints. Imagine the ends of the members as fixed and find the fixed end moments at the joints. Equivalent joint loads are got by reversing the sign of fixed end moments.

In this problem, following are the fixed end moments (FEMs)

$$MF_{AB} = -\frac{wl^2}{12} = -90 \text{ kN-m}$$

$$MF_{BA} = \frac{wl^2}{12} = 90 \text{ kN-m}$$

$$MF_{BC} = -\frac{wl}{8} = -40 \text{ kN-m}$$

$$MF_{CB} = \frac{wl}{8} = 40 \text{ kN-m}$$

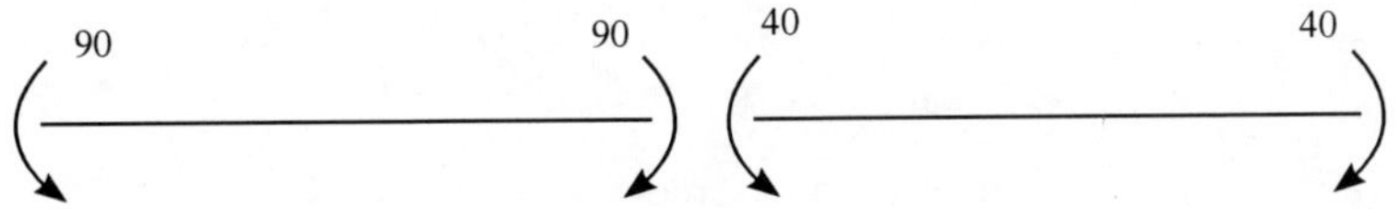

Fixed end moments

Fig. 2.4(d)

For the given beam, the moment MF_{AB} can be ignored since A is fixed. At support B, we have a clockwise moment of 90 kN-m in span BA and anticlockwise moment of 40 kN-m in span BC. For the equilibrium of joint B, we have to apply anticlockwise moment of 50 kN-m in span BC.

Support C is simply supported end. We imagined the ends of the members as fixed and found the fixed end moments at the joints. But, support C is simply supported end. Therefore, apply an anticlockwise moment of 40 kN-m.

Here, anticlockwise moment of 50 kN-m at support B and anticlockwise moment of 40 kN-m at support C are the forces resulting from the given external forces as shown in Fig. 2.4(e).

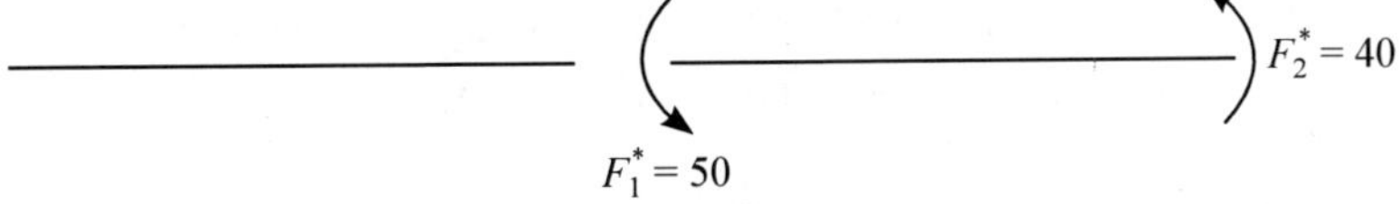

System coordinates

Fig. 2.4(e)

Hence, $\{F^*\} = \begin{Bmatrix} F_1^* \\ F_2^* \end{Bmatrix} = \begin{Bmatrix} 50 \\ 40 \end{Bmatrix}$

where, $\{F^*\}$= system forces or system coordinates.

Now the force vector, $\{F\}$ comprises two classes of loads, namely, the applied forces or the system forces, $\{F^*\}$ and the redundant forces, $\{F^0\}$.

$$\{F\} = \begin{Bmatrix} F_1^* \\ F_2^* \\ F_3^0 \\ F_4^0 \end{Bmatrix}$$

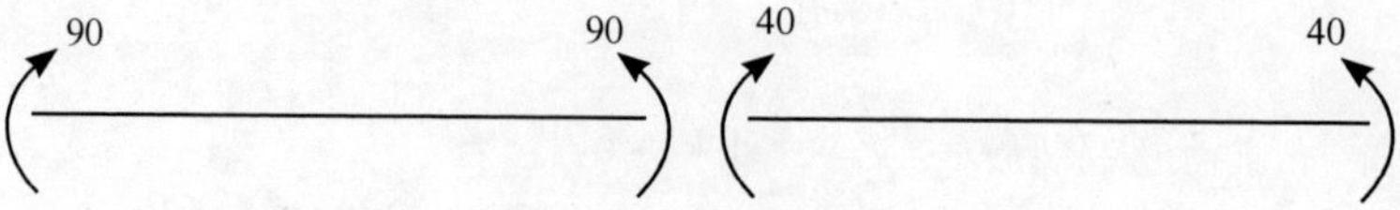

Equivalent joint forces

Fig. 2.4(f)

$$\{P^e\} = \begin{Bmatrix} 90 \\ -90 \\ 40 \\ -40 \end{Bmatrix}$$

where, $\{P^e\}$ = vector of equivalent joint forces.

Step 4:

Develop the transformation matrix, $[b]$

Transformation matrix is defined as the matrix connecting vectors in local coordinate system and global coordinate system. The flexibility matrices developed for the individual elements have to be transformed to global coordinates for which transformation matrix is used.

Generate the $[b]$ matrix such that $\{P\} = [b]\ \{F\}$.

In the given problem, the $[b]$ matrix is generated by applying $F_1^* = 1$, $F_2^* = 1$, $F_3^0 = 1$ and $F_4^0 = 1$ in different steps.

For getting the first column of the $[b]$ matrix, apply unit force along F_1^*.

Similarly, to get the second column of the $[b]$ matrix, apply unit force along F_2^*.

Similarly, to get the third column of the $[b]$ matrix, apply unit force along F_3^0.

And to get the fourth column of the $[b]$ matrix, apply unit force along F_4^0 as shown in Fig. 2.4(g).

Note: Compare Fig. 2.4(g), with the element coordinates (Refer Fig. 2.4(c)) and then fill the $[b]$ - matrix.

$$[b] = \left[\begin{array}{cc:cc} 0 & 0 & 1 & 0 \\ 0 & 0 & 0 & -1 \\ -1 & 0 & 0 & 1 \\ 0 & -1 & 0 & 0 \end{array}\right]$$
$$\qquad\quad b^* \qquad\quad b^0$$

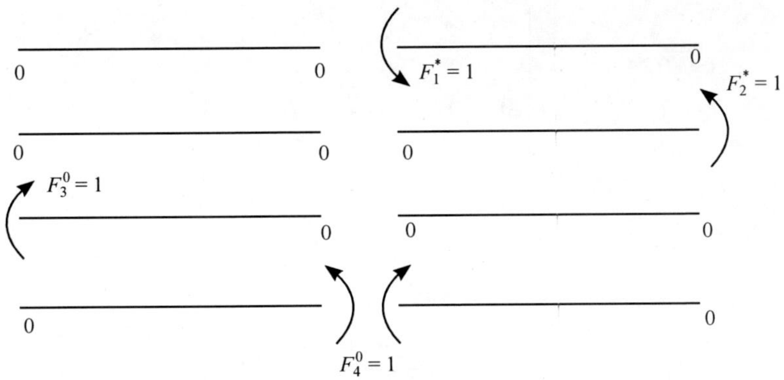

Fig. 2.4(g)

The part of the [b] matrix relating to F_1^* and F_2^* is called b^* and the other part corresponding to F_3^0 and F_4^0 is called b^0.

Step 5:

Synthesize the element flexibility matrix [α]

In the given problem,

For element 1,

$$[\alpha_1] = \frac{l_1}{6E_1I_1}\begin{bmatrix} 2 & -1 \\ -1 & 2 \end{bmatrix} = \frac{6}{6EI}\begin{bmatrix} 2 & -1 \\ -1 & 2 \end{bmatrix} = \frac{1}{EI}\begin{bmatrix} 2 & -1 \\ -1 & 2 \end{bmatrix}$$

For element 2,

$$[\alpha_2] = \frac{l_2}{6E_2I_2}\begin{bmatrix} 2 & -1 \\ -1 & 2 \end{bmatrix} = \frac{4}{6EI}\begin{bmatrix} 2 & -1 \\ -1 & 2 \end{bmatrix} = \frac{1}{EI}\begin{bmatrix} 1.33 & -0.67 \\ -0.67 & 1.33 \end{bmatrix}$$

Assembled element flexibility matrix, [α]

$$[\alpha] = \begin{bmatrix} [\alpha_1] & 0 \\ 0 & [\alpha_2] \end{bmatrix}$$

$$[\alpha] = \frac{1}{EI}\begin{bmatrix} 2 & -1 & 0 & 0 \\ -1 & 2 & 0 & 0 \\ 0 & 0 & 1.33 & -0.67 \\ 0 & 0 & -0.67 & 1.33 \end{bmatrix}$$

Step 6:

Compute the following, system flexibility matrix, [a]

$$[a] = [b]^T[\alpha][b]$$

$$[a^{00}] = [b^0]^T[\alpha][b^0]$$

$$[a^{00}] = \begin{bmatrix} 1 & 0 & 0 & 0 \\ 0 & -1 & 1 & 0 \end{bmatrix} \times \frac{1}{EI}\begin{bmatrix} 2 & -1 & 0 & 0 \\ -1 & 2 & 0 & 0 \\ 0 & 0 & 1.33 & -0.67 \\ 0 & 0 & -0.67 & 1.33 \end{bmatrix} \times \begin{bmatrix} 1 & 0 \\ 0 & -1 \\ 0 & 1 \\ 0 & 0 \end{bmatrix}$$

$$[a^{00}] = \frac{1}{EI}\begin{bmatrix} 2 & 1 \\ 1 & 3.33 \end{bmatrix}$$

$$[a^{00}]^{-1} = \frac{EI}{5.66}\begin{bmatrix} 3.33 & -1 \\ -1 & 2 \end{bmatrix}$$

$$[a^{0*}] = [b^0]^T[\alpha][b^*]$$

$$[a^{0*}] = \begin{bmatrix} 1 & 0 & 0 & 0 \\ 0 & -1 & 1 & 0 \end{bmatrix} \times \frac{1}{EI}\begin{bmatrix} 2 & -1 & 0 & 0 \\ -1 & 2 & 0 & 0 \\ 0 & 0 & 1.33 & -0.67 \\ 0 & 0 & -0.67 & 1.33 \end{bmatrix} \times \begin{bmatrix} 0 & 0 \\ 0 & 0 \\ -1 & 0 \\ 0 & -1 \end{bmatrix}$$

$$[a^{0*}] = \frac{1}{EI}\begin{bmatrix} 0 & 0 \\ -1.33 & 0.67 \end{bmatrix}$$

Compute the Redundant Forces

$$\{F^0\} = -[a^{00}]^{-1}[a^{0*}]\{F^*\}$$

$$\{F^*\} = \begin{Bmatrix} 50 \\ 40 \end{Bmatrix}$$

$$\{F^0\} = -[a^{00}]^{-1}[a^{0*}]\{F^*\}$$

$$\{F^0\} = -\frac{EI}{5.66}\begin{bmatrix} 3.33 & -1 \\ -1 & 2 \end{bmatrix} \times \frac{1}{EI}\begin{bmatrix} 0 & 0 \\ -1.33 & 0.67 \end{bmatrix} \times \begin{Bmatrix} 50 \\ 40 \end{Bmatrix}$$

$$\{F^0\} = \begin{Bmatrix} -7.01 \\ 14.03 \end{Bmatrix} = \begin{Bmatrix} F_3^0 \\ F_4^0 \end{Bmatrix} \text{ kN-m}$$

Step 7:

Compute element forces

$$\{P\} = [b]\{F\}$$

$$\begin{Bmatrix} P_1 \\ P_2 \\ P_3 \\ P_4 \end{Bmatrix} = \begin{bmatrix} 0 & 0 & 1 & 0 \\ 0 & 0 & 0 & -1 \\ -1 & 0 & 0 & 1 \\ 0 & -1 & 0 & 0 \end{bmatrix} \times \begin{Bmatrix} 50 \\ 40 \\ -7.01 \\ 14.03 \end{Bmatrix} = \begin{Bmatrix} -7.01 \\ -14.03 \\ -35.97 \\ -40 \end{Bmatrix}$$

Step 8:

Compute final forces

$\{P^f\} = \{P\} - \{P^e\}$

$$\{P^f\} = \begin{Bmatrix} -7.01 \\ -14.03 \\ -35.97 \\ -40 \end{Bmatrix} - \begin{Bmatrix} 90 \\ -90 \\ 40 \\ -40 \end{Bmatrix} = \begin{Bmatrix} -97.07 \\ 75.97 \\ -75.97 \\ 0 \end{Bmatrix}$$

Final moments and bending moment diagrams as shown in Fig. 2.4(h) and (i)

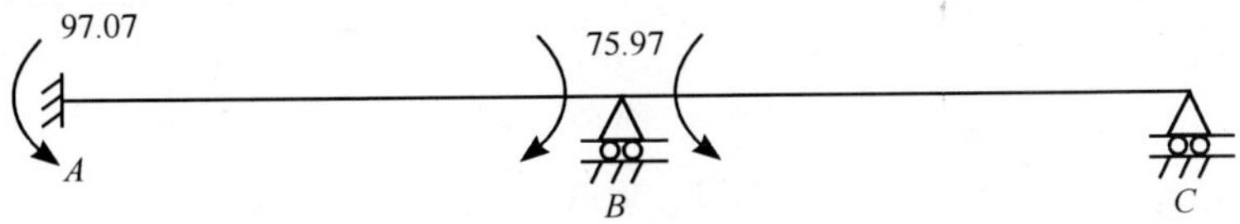

End moments

Fig. 2.4(h)

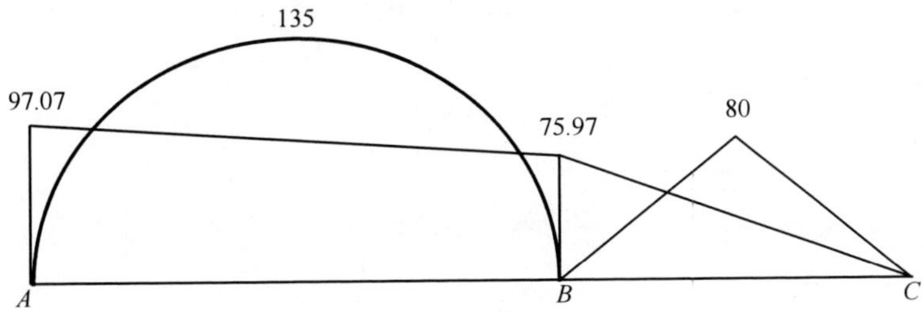

Final BMD

Fig. 2.4(i)

Example 2.3 Analyze the continuous beam shown in Fig. 2.5(a) flexibility matrix method.

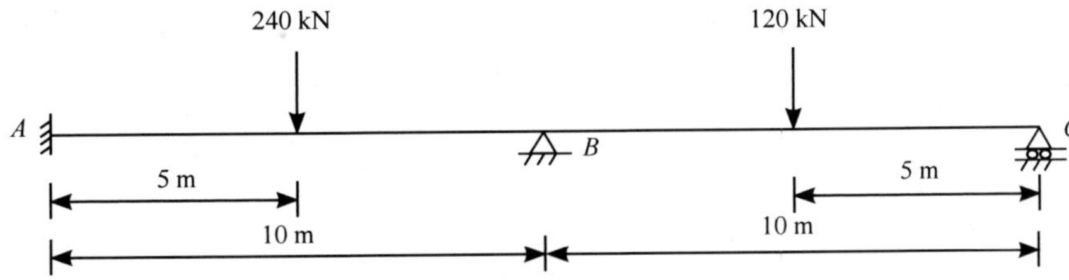

Fig. 2.5(a)

Solution:

The degree of static indeterminacy = 2

The primary structure is chosen as below, by introducing hinges at A and B. The moments at A and B are thus redundants as shown in Fig. 2.5(b).

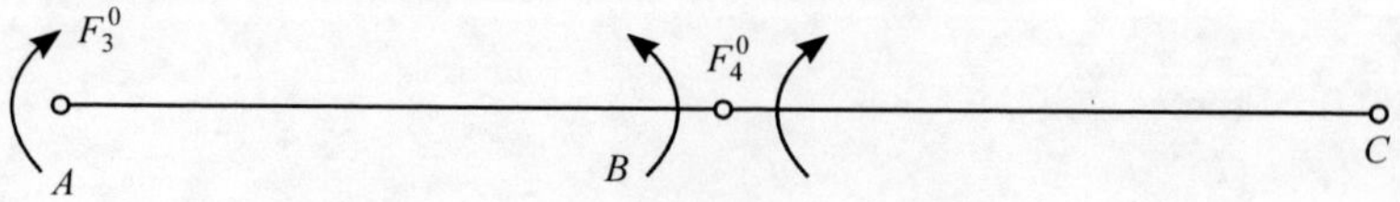

Primary structure chosen by choosing hinges at *A* and *B*.

Fig. 2.5(b)

Select {*P*} coordinates as shown in Fig. 2.5(c).

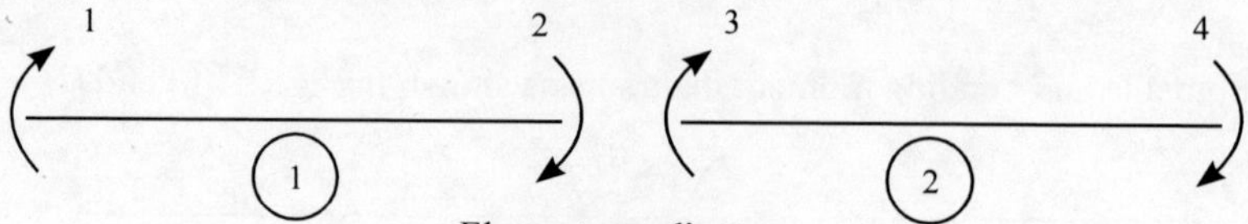

Element coordinates

Fig. 2.5(c)

Finding fixed end moments (FEMs)

$$MF_{AB} = -\frac{wl}{8} = -300 \text{ kN-m}$$

$$MF_{BA} = \frac{wl}{8} = 300 \text{ kN-m}$$

$$MF_{BC} = -\frac{wl}{8} = -150 \text{ kN-m}$$

$$MF_{CB} = \frac{wl}{8} = 150 \text{ kN-m}$$

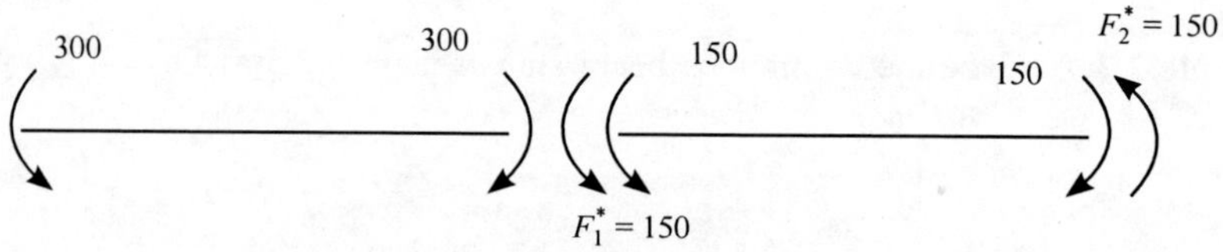

FEMs & system coordinates

Fig. 2.5(d)

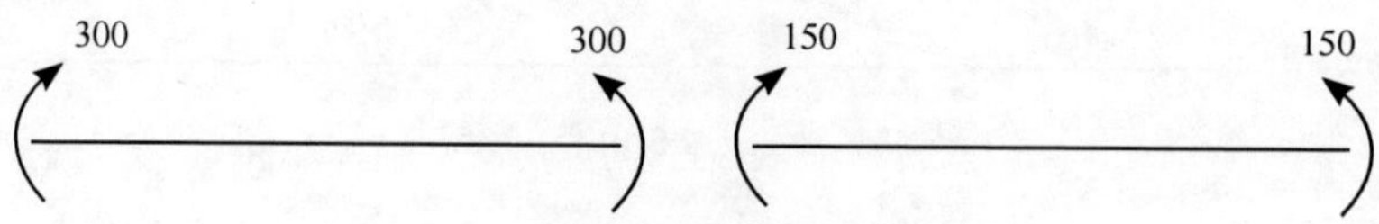

Equivalent joint forces

Fig. 2.5(e)

The $[b]$ – matrix is generated by introducing $F_1^* = 1$, $F_2^* = 1$, $F_3^0 = 1$ and $F_4^0 = 1$ in different steps.

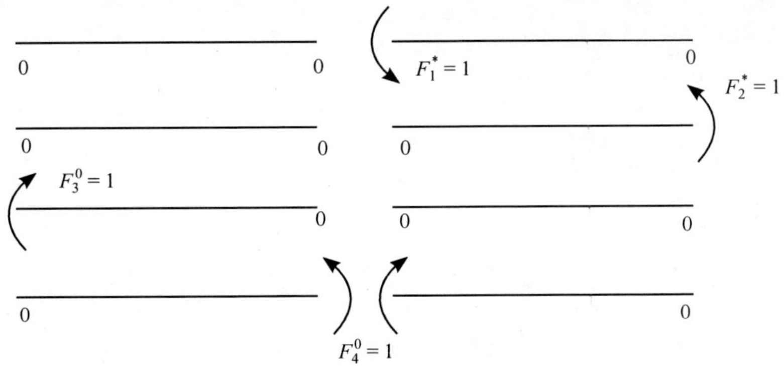

Fig. 2.5(f)

$$[b] = \underbrace{\begin{bmatrix} 0 & 0 \\ 0 & 0 \\ -1 & 0 \\ 0 & -1 \end{bmatrix}}_{b^*} \vdots \underbrace{\begin{bmatrix} 1 & 0 \\ 0 & -1 \\ 0 & 1 \\ 0 & 0 \end{bmatrix}}_{b^0}$$

Element flexibility matrix for element 1 and 2 is

$$[\alpha_1] = \frac{l_1}{6E_1I_1}\begin{bmatrix} 2 & -1 \\ -1 & 2 \end{bmatrix} = \frac{10}{6EI}\begin{bmatrix} 2 & -1 \\ -1 & 2 \end{bmatrix} = [\alpha_2]$$

$$[\alpha] = \begin{bmatrix} \alpha_1 & 0 \\ 0 & \alpha_2 \end{bmatrix}$$

Assembled element flexibility matrix, $[\alpha]$ is

$$[\alpha] = \frac{10}{6EI}\begin{bmatrix} 2 & -1 & 0 & 0 \\ -1 & 2 & 0 & 0 \\ 0 & 0 & 2 & -1 \\ 0 & 0 & -1 & 2 \end{bmatrix}$$

$$[a] = [b]^T[\alpha][b]$$

$$[a^{00}] = [b^0]^T[\alpha][b^0]$$

$$[a^{00}] = \begin{bmatrix} 1 & 0 & 0 & 0 \\ 0 & -1 & 1 & 0 \end{bmatrix} \times \frac{10}{6EI}\begin{bmatrix} 2 & -1 & 0 & 0 \\ -1 & 2 & 0 & 0 \\ 0 & 0 & 2 & -1 \\ 0 & 0 & -1 & 2 \end{bmatrix} \times \begin{bmatrix} 1 & 0 \\ 0 & -1 \\ 0 & 1 \\ 0 & 0 \end{bmatrix}$$

$$[a^{00}] = \frac{10}{6EI}\begin{bmatrix} 2 & 1 \\ 1 & 4 \end{bmatrix}$$

$$[a^{00}]^{-1} = \frac{6EI}{10} \times \frac{1}{7} \times \begin{bmatrix} 4 & -1 \\ -1 & 2 \end{bmatrix}$$

$$[a^{0*}] = [b^0]^T[\alpha][b^*]$$

$$[a^{0*}] = \frac{10}{6EI}\begin{bmatrix} 0 & 0 \\ -2 & 1 \end{bmatrix}$$

Redundant forces are given by

$$\{F^0\} = -[a^{00}]^{-1}[a^{0*}]\{F^*\}$$

$$\{F^0\} = \begin{Bmatrix} -21.43 \\ 42.86 \end{Bmatrix} = \begin{Bmatrix} F_3^0 \\ F_4^0 \end{Bmatrix}$$

The element forces are given by

$$\{P\} = [b]\,\{F\}$$

$$\begin{Bmatrix} P_1 \\ P_2 \\ P_3 \\ P_4 \end{Bmatrix} = \begin{bmatrix} 0 & 0 & 1 & 0 \\ 0 & 0 & 0 & -1 \\ -1 & 0 & 0 & 1 \\ 0 & -1 & 0 & 0 \end{bmatrix} \times \begin{Bmatrix} 150 \\ 150 \\ -21.43 \\ 42.86 \end{Bmatrix} = \begin{Bmatrix} -21.43 \\ -42.86 \\ -107.14 \\ -150 \end{Bmatrix}$$

The final forces are given by

$$\{P^f\} = \{P\} - \{P^e\} = \begin{Bmatrix} -21.43 \\ -42.86 \\ -107.14 \\ -150 \end{Bmatrix} - \begin{Bmatrix} 300 \\ -300 \\ 150 \\ -150 \end{Bmatrix} = \begin{Bmatrix} -321.83 \\ 257.14 \\ -257.14 \\ 0 \end{Bmatrix}$$

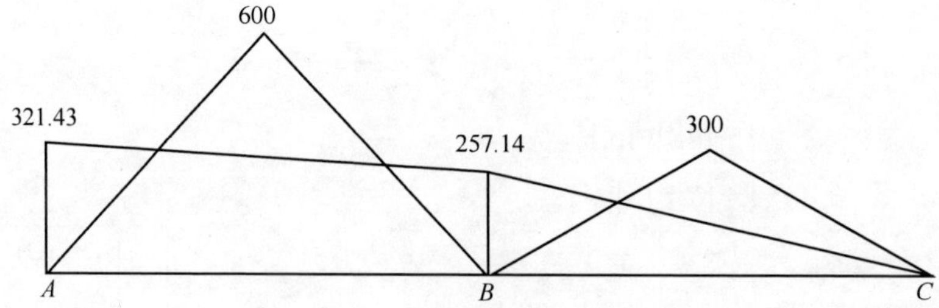

Final BMD

Fig. 2.5(g)

Example 2.4 Analyze the continuous beam shown in Fig. 2.6(a) by flexibility matrix method.

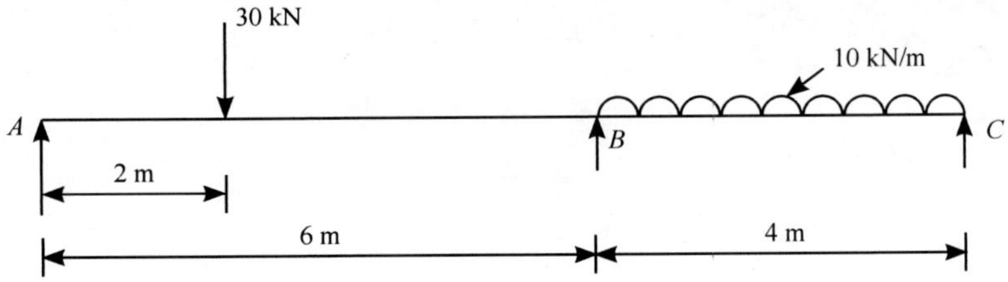

Fig. 2.6(a)

Solution:

Static indeterminacy = 1

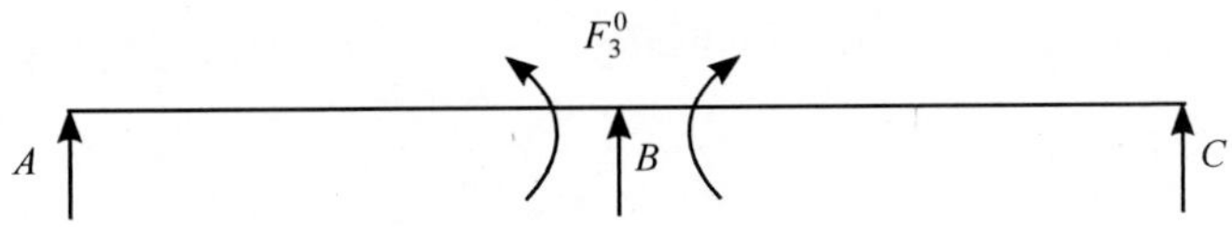

Primary structure with redundant F^0

Fig. 2.6(b)

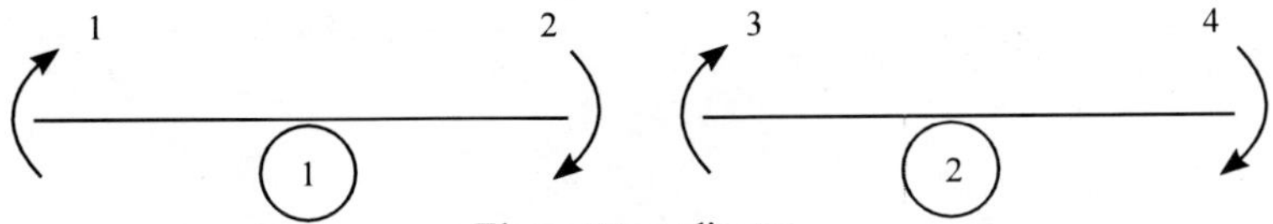

Element coordinates

Fig. 2.6(c)

Finding fixed end moments (FEMs)

$$MF_{AB} = -\frac{wab^2}{l^2} = -26.67 \text{ kN-m}$$

$$MF_{BA} = \frac{wa^2b}{l^2} = 13.33 \text{ kN-m}$$

$$MF_{BC} = -\frac{wl^2}{12} = -13.33 \text{ kN-m}$$

$$MF_{CB} = \frac{wl^2}{12} = 13.33 \text{ kN-m}$$

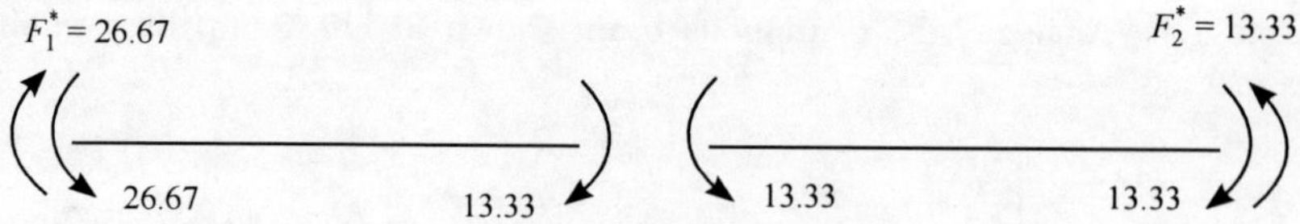

FEMs & system coordinates

Fig. 2.6(d)

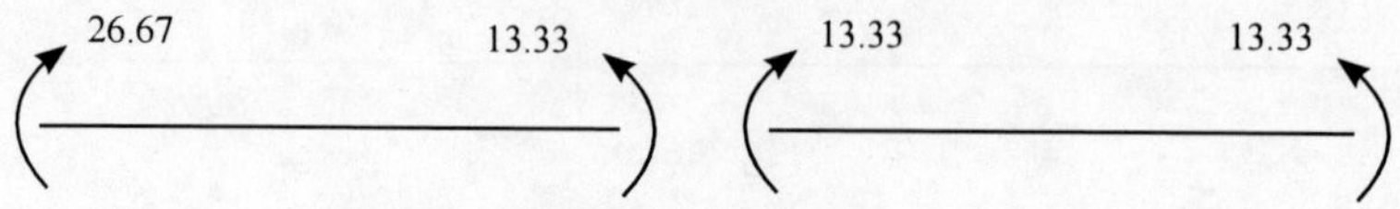

Equivalent joint forces

Fig. 2.6(e)

The $[b]$ – matrix is generated by introducing $F_1^* = 1$, $F_2^* = 1$ and $F_3^0 = 1$ in different steps.

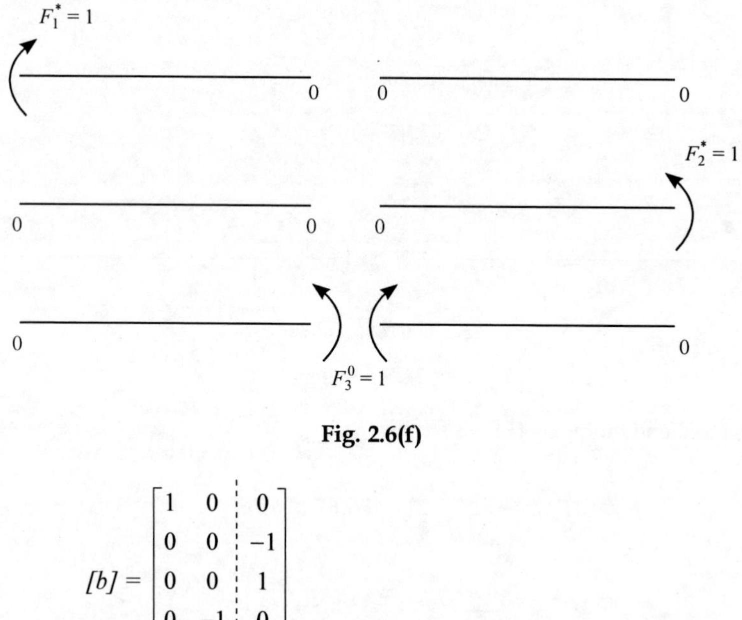

Fig. 2.6(f)

$$[b] = \left[\begin{array}{cc:c} 1 & 0 & 0 \\ 0 & 0 & -1 \\ 0 & 0 & 1 \\ 0 & -1 & 0 \end{array}\right]$$
$$\qquad\quad b^* \qquad\quad b^0$$

Element flexibility matrix for element 1 and 2 is

$$[\alpha_1] = \frac{6}{6EI}\begin{bmatrix} 2 & -1 \\ -1 & 2 \end{bmatrix} = \frac{1}{EI}\begin{bmatrix} 2 & -1 \\ -1 & 2 \end{bmatrix}$$

$$[\alpha_2] = \frac{4}{6EI}\begin{bmatrix} 2 & -1 \\ -1 & 2 \end{bmatrix} = \frac{1}{EI}\begin{bmatrix} 1.33 & -0.67 \\ -0.67 & 1.33 \end{bmatrix}$$

$$[\alpha] = \begin{bmatrix} \alpha_1 & 0 \\ 0 & \alpha_2 \end{bmatrix}$$

Assembled element flexibility matrix, [α] is

$$[\alpha] = \frac{1}{EI}\begin{bmatrix} 2 & -1 & 0 & 0 \\ -1 & 2 & 0 & 0 \\ 0 & 0 & 1.33 & -0.67 \\ 0 & 0 & -0.67 & 1.33 \end{bmatrix}$$

$$[a^{00}] = [b^0]^T[\alpha][b^0]$$

$$[a^{00}] = \begin{bmatrix} 0 & -1 & 1 & 0 \end{bmatrix} \times \frac{1}{EI}\begin{bmatrix} 2 & -1 & 0 & 0 \\ -1 & 2 & 0 & 0 \\ 0 & 0 & 1.33 & -0.67 \\ 0 & 0 & -0.67 & 1.33 \end{bmatrix} \times \begin{bmatrix} 0 \\ -1 \\ 1 \\ 0 \end{bmatrix}$$

$$[a^{00}] = \frac{1}{EI}\begin{bmatrix} 3.33 \end{bmatrix}$$

$$[a^{00}]^{-1} = \frac{EI}{3.33}$$

$$[a^{0*}] = [b^0]^T[\alpha][b^*]$$

$$[a^{0*}] = \begin{bmatrix} 0 & -1 & 1 & 0 \end{bmatrix} \times \frac{1}{EI}\begin{bmatrix} 2 & -1 & 0 & 0 \\ -1 & 2 & 0 & 0 \\ 0 & 0 & 1.33 & -0.67 \\ 0 & 0 & -0.67 & 1.33 \end{bmatrix} \times \begin{bmatrix} 1 & 0 \\ 0 & 0 \\ 0 & 0 \\ 0 & -1 \end{bmatrix}$$

$$[a^{0*}] = \frac{1}{EI}\begin{bmatrix} 1 & 0.67 \end{bmatrix}$$

Redundant forces are given by

$$\{F^0\} = -[a^{00}]^{-1}[a^{0*}]\{F^*\}$$

$$\{F^*\} = \begin{Bmatrix} 26.67 \\ 13.33 \end{Bmatrix}$$

$$\{F^0\} = -10.69 \text{ kN-m}$$

Element forces are given by

$$\{P\} = [b]\{F\}$$

$$\begin{Bmatrix} P_1 \\ P_2 \\ P_3 \\ P_4 \end{Bmatrix} = \begin{bmatrix} 1 & 0 & 0 \\ 0 & 0 & -1 \\ 0 & 0 & 1 \\ 0 & -1 & 0 \end{bmatrix} \times \begin{Bmatrix} 26.67 \\ 13.33 \\ -10.69 \end{Bmatrix} = \begin{Bmatrix} 26.67 \\ 10.69 \\ -10.69 \\ -13.33 \end{Bmatrix}$$

Final forces are given by

$$\{P^f\} = \{P\} - \{P^e\}$$

$$= \begin{Bmatrix} 26.67 \\ 10.69 \\ -10.69 \\ -13.33 \end{Bmatrix} - \begin{Bmatrix} 26.67 \\ -13.33 \\ 13.33 \\ -13.33 \end{Bmatrix} = \begin{Bmatrix} 0 \\ 24.02 \\ -24.02 \\ 0 \end{Bmatrix}$$

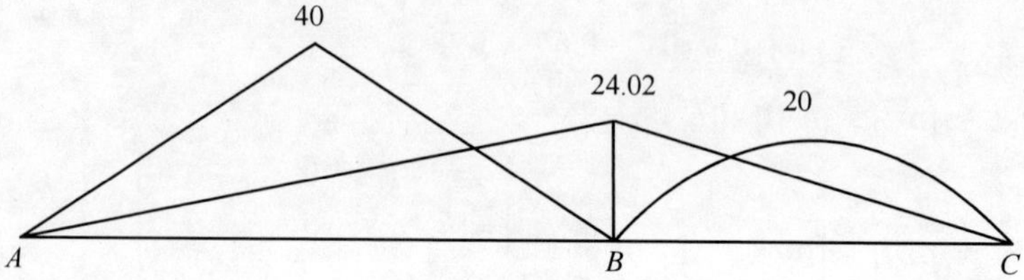

Final BMD

Fig. 2.6(g)

Example 2.5 Analyze the continous beam shown in Fig. 2.7(a) by flexibility matrix method.

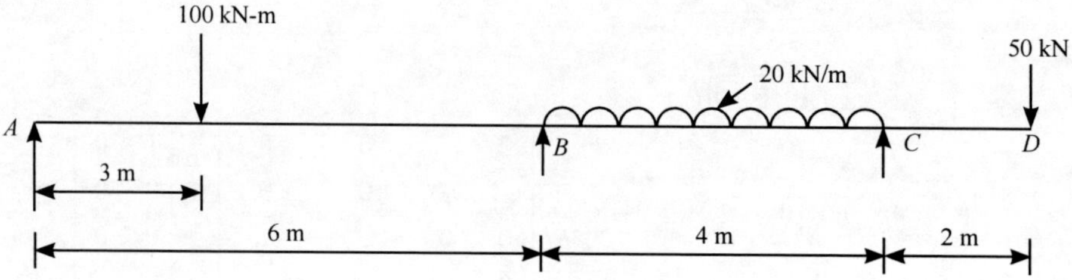

Fig. 2.7(a)

Solution:

Static indeterminacy = 2

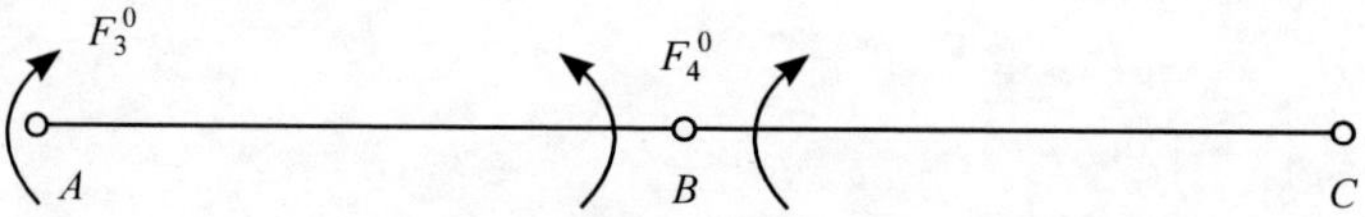

Primary structure with redundants

$F_2^* = 1$

Fig. 2.7(b)

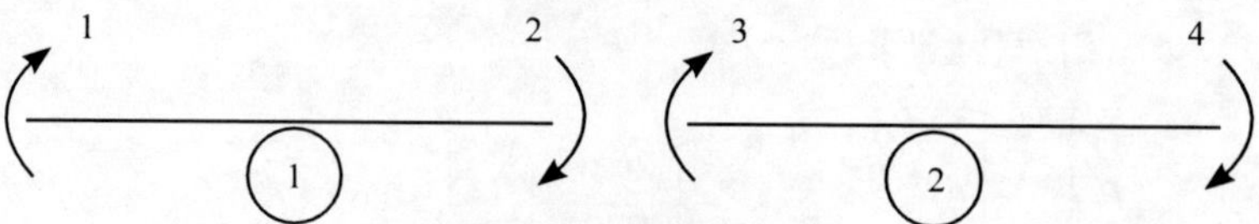

Element coordinates

Fig. 2.7(c)

Finding fixed end moments (FEMs)

$$MF_{AB} = -\frac{M}{4} = -25 \text{ kN-m}$$

$$MF_{BA} = -\frac{M}{4} = -25 \text{ kN-m}$$

$$MF_{BC} = -\frac{wl^2}{12} = -60 \text{ kN-m}$$

$$MF_{CB} = \frac{wl^2}{12} = 60 \text{ kN-m}$$

$$M_{CD} = -50 \times 2 = -100 \text{ kN-m}$$

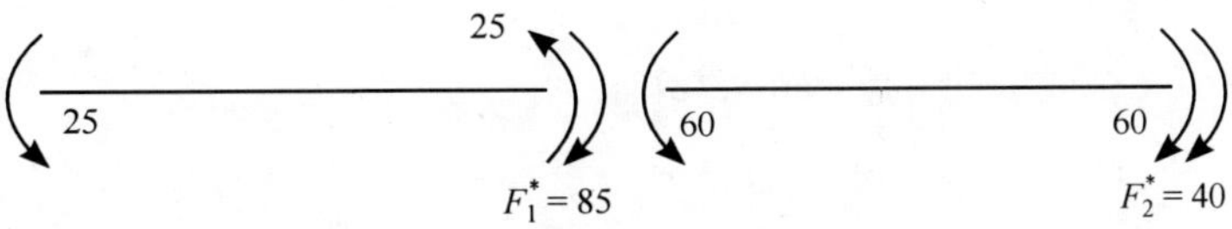

FEMs system coordinates

Fig. 2.7(d)

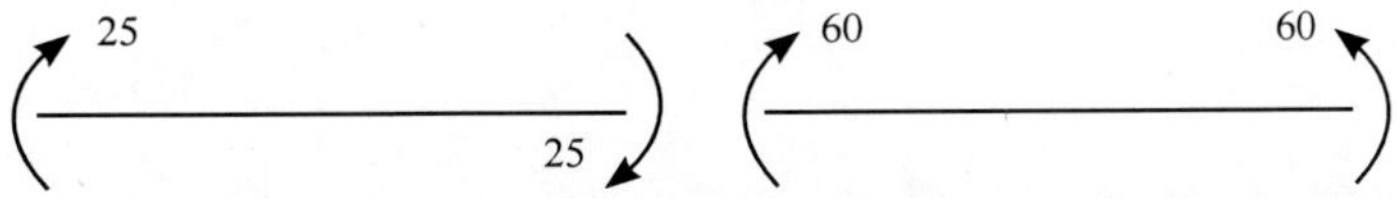

Equivalent joint forces

Fig. 2.7(e)

The $[b]$ – matrix is generated by introducing $F_1^* = 1$, $F_2^* = 1$, $F_3^0 = 1$ and $F_4^0 = 1$ in different steps.

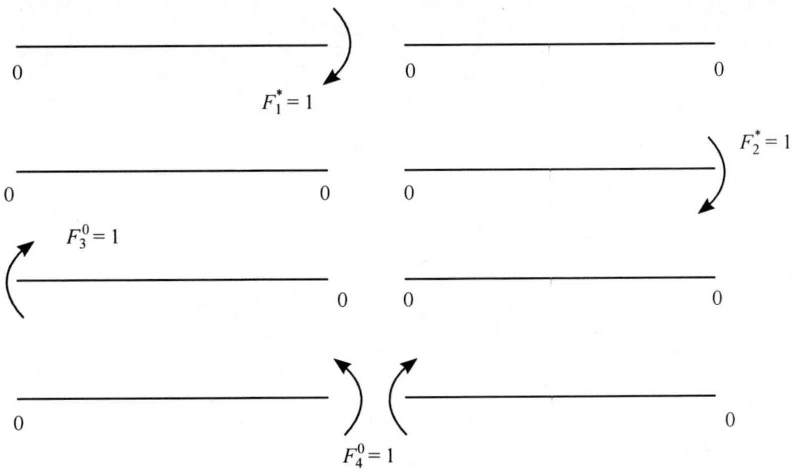

Fig. 2.7(f)

$$[b] = \left[\begin{array}{cc:cc} 0 & 0 & 1 & 0 \\ 1 & 0 & 0 & -1 \\ 0 & 0 & 0 & 1 \\ 0 & 1 & 0 & 0 \end{array}\right]$$
$$\quad b^* \qquad\qquad b^0$$

Element flexibility matrix element 1 and 2 is

$$[\alpha_1] = \frac{l_1}{6E_1I_1}\begin{bmatrix} 2 & -1 \\ -1 & 2 \end{bmatrix} = \frac{1}{EI}\begin{bmatrix} 2 & -1 \\ -1 & 2 \end{bmatrix} = [\alpha_2]$$

$$[\alpha] = \begin{bmatrix} \alpha_1 & 0 \\ 0 & \alpha_2 \end{bmatrix}$$

Assembled element flexibility matrix, $[\alpha]$ is

$$[\alpha] = \frac{1}{EI}\begin{bmatrix} 2 & -1 & 0 & 0 \\ -1 & 2 & 0 & 0 \\ 0 & 0 & 2 & -1 \\ 0 & 0 & -1 & 2 \end{bmatrix}$$

$$[a^{00}] = [b^0]^T[\alpha][b^0]$$

$$[a^{00}] = \begin{bmatrix} 1 & 0 & 0 & 0 \\ 0 & -1 & 1 & 0 \end{bmatrix} \times \frac{1}{EI}\begin{bmatrix} 2 & -1 & 0 & 0 \\ -1 & 2 & 0 & 0 \\ 0 & 0 & 2 & -1 \\ 0 & 0 & -1 & 2 \end{bmatrix} \times \begin{bmatrix} 1 & 0 \\ 0 & -1 \\ 0 & 1 \\ 0 & 0 \end{bmatrix}$$

$$[a^{00}] = \frac{1}{EI}\begin{bmatrix} 2 & 1 \\ 1 & 4 \end{bmatrix}$$

$$[a^{00}]^{-1} = EI \times \frac{1}{7} \times \begin{bmatrix} 4 & -1 \\ -1 & 2 \end{bmatrix}$$

$$[a^{0*}] = [b^0]^T[\alpha][b^*]$$

$$[a^{0*}] = \begin{bmatrix} 1 & 0 & 0 & 0 \\ 0 & -1 & 1 & 0 \end{bmatrix} \times \frac{1}{EI}\begin{bmatrix} 2 & -1 & 0 & 0 \\ -1 & 2 & 0 & 0 \\ 0 & 0 & 2 & -1 \\ 0 & 0 & -1 & 2 \end{bmatrix} \times \begin{bmatrix} 0 & 0 \\ 1 & 0 \\ 0 & 0 \\ 0 & 1 \end{bmatrix}$$

$$[a^{0*}] = \frac{1}{EI}\begin{bmatrix} -1 & 0 \\ -2 & -1 \end{bmatrix}$$

Redundant forces are given by

$$\{F^0\} = -[a^{00}]^{-1}[a^{0*}]\{F^*\}$$

$$\{F^*\} = \begin{Bmatrix} 85 \\ 40 \end{Bmatrix}$$

$$\{F^0\} = \begin{Bmatrix} 18.57 \\ 47.86 \end{Bmatrix} = \begin{Bmatrix} F_3^0 \\ F_4^0 \end{Bmatrix}$$

Element forces are given by

$$\{P\} = [b]\{F\}$$

$$\begin{Bmatrix} P_1 \\ P_2 \\ P_3 \\ P_4 \end{Bmatrix} = \begin{bmatrix} 0 & 0 & 1 & 0 \\ 1 & 0 & 0 & -1 \\ 0 & 0 & 0 & 1 \\ 0 & 1 & 0 & 0 \end{bmatrix} \times \begin{Bmatrix} 85 \\ 40 \\ 18.57 \\ 47.86 \end{Bmatrix} = \begin{Bmatrix} 18.57 \\ 37.14 \\ 47.86 \\ 40 \end{Bmatrix}$$

Final forces are given by

$$\{P^f\} = \{P\} - \{P^e\}$$

$$\{P^f\} = \begin{Bmatrix} 18.57 \\ 37.14 \\ 47.86 \\ 40 \end{Bmatrix} - \begin{Bmatrix} 25 \\ 25 \\ 60 \\ -60 \end{Bmatrix} = \begin{Bmatrix} -6.43 \\ 12.14 \\ -12.14 \\ 100 \end{Bmatrix}$$

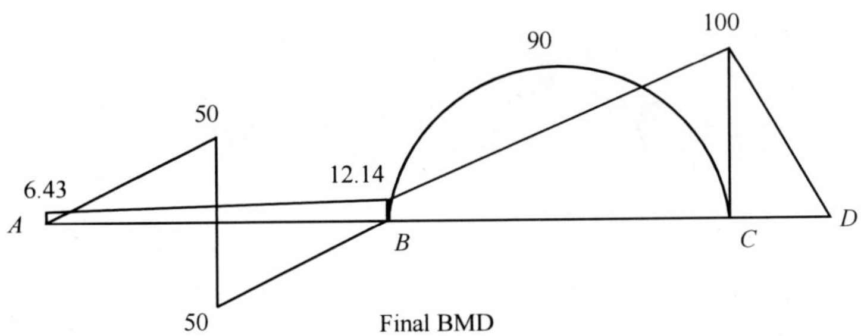

Fig. 2.7(g)

2.3 ANALYSIS OF PLANE RIGID FRAMES

Plane rigid frames are analysed here by flexibility matrix method.

Example 2.6 Analyze the given frame as shown in Fig. 2.8(a) by matrix flexibility method.

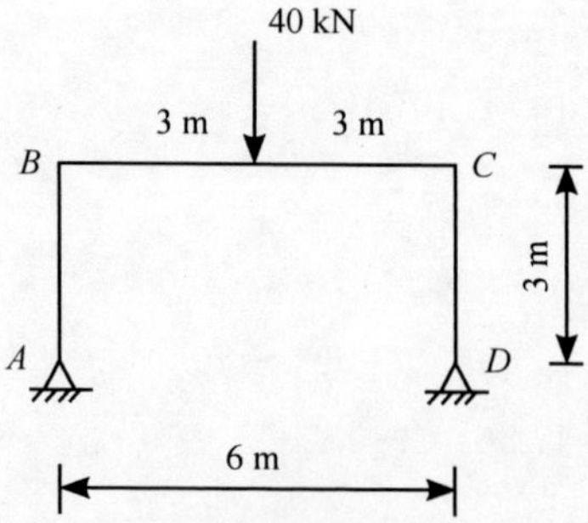

Fig. 2.8(a)

Solution:

Static indeterminacy = 1

Treating the horizontal reaction at D as the redundant, the primary structure will be as shown in Fig. 2.8(b).

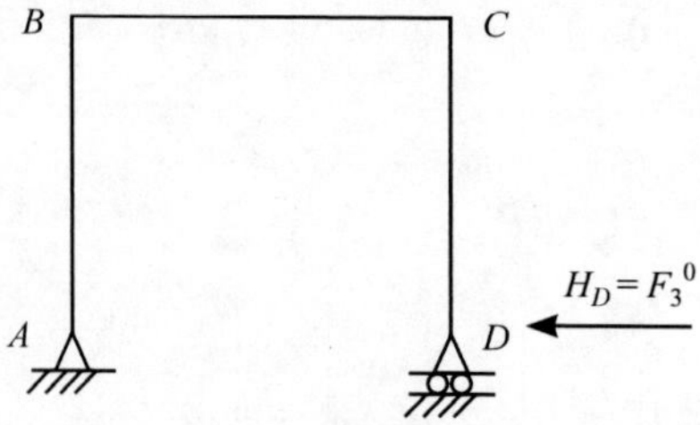

Fig. 2.8(b)

Fixed end moments

$$MF_{AB} = MF_{BA} = MF_{CD} = MF_{DC} = 0$$

$$MF_{BC} = -\frac{wl}{8} = -30 \text{ kN-m}$$

$$MF_{CB} = \frac{wl}{8} = 30 \text{ kN-m}$$

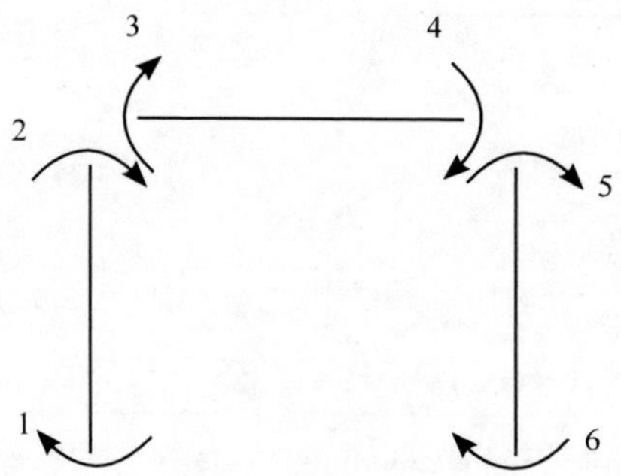

Fig. 2.8(c)

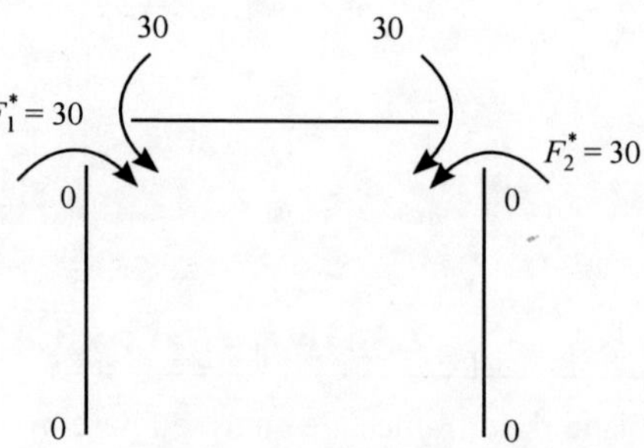

Fig. 2.8(d)

Formation of [b] – Matrix

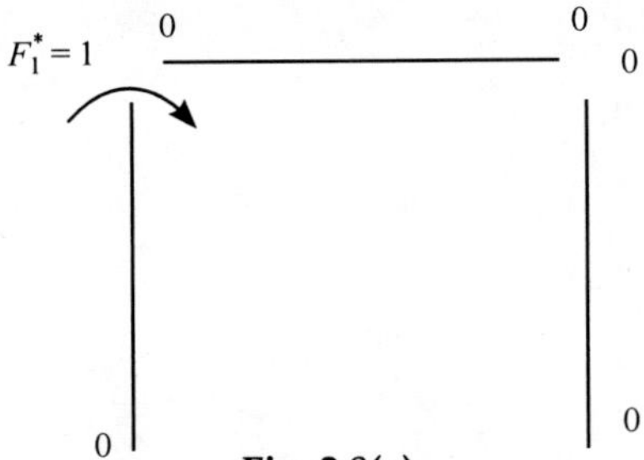

Fig. 2.8(e)

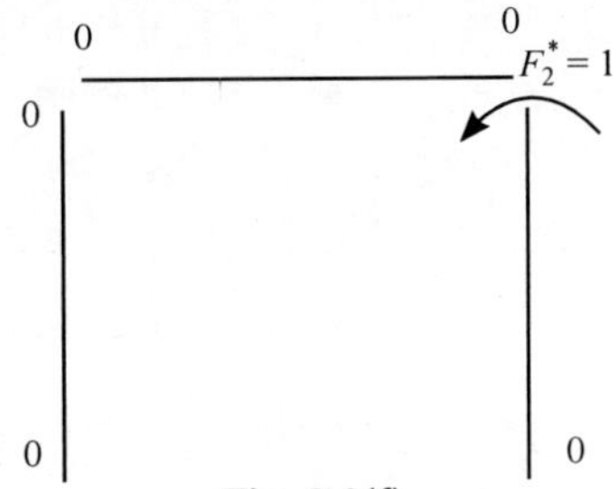

Fig. 2.8(f)

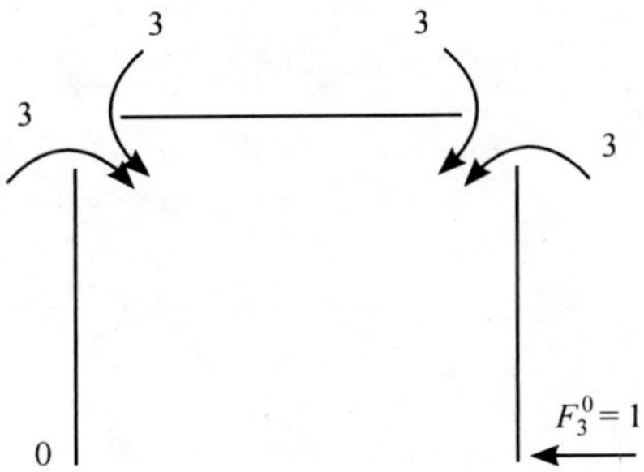

Fig. 2.8(g)

$$[b] = \left[\begin{array}{cc:c} 0 & 0 & 0 \\ 1 & 0 & 3 \\ 0 & 0 & -3 \\ 0 & 0 & 3 \\ 0 & -1 & -3 \\ 0 & 0 & 0 \end{array}\right]$$

$$\quad b^* \qquad\quad b^0$$

Element Flexibility Matrix

$$[\alpha_1] = \frac{l_1}{6E_1I_1}\begin{bmatrix} 2 & -1 \\ -1 & 2 \end{bmatrix} = \frac{1}{2EI}\begin{bmatrix} 2 & -1 \\ -1 & 2 \end{bmatrix} = [\alpha_3]$$

$$[\alpha_2] = \frac{l_2}{6E_2I_2}\begin{bmatrix} 2 & -1 \\ -1 & 2 \end{bmatrix} = \frac{}{EI}\begin{bmatrix} 4 & 2 \\ 2 & 4 \end{bmatrix}$$

Assembled Element Flexibility Matrix

$$[\alpha] = \frac{1}{2EI}\begin{bmatrix} 2 & -1 & 0 & 0 & 0 & 0 \\ -1 & 2 & 0 & 0 & 0 & 0 \\ 0 & 0 & 4 & -2 & 0 & 0 \\ 0 & 0 & -2 & 4 & 0 & 0 \\ 0 & 0 & 0 & 0 & 2 & -1 \\ 0 & 0 & 0 & 0 & -1 & 2 \end{bmatrix}$$

$$[a^{00}] = [b^0]^T[\alpha][b^0]$$

$$[a^{00}] = \begin{bmatrix} 0 & 3 & -3 & 3 & -3 & 0 \end{bmatrix} \times \frac{1}{2EI}\begin{bmatrix} 2 & -1 & 0 & 0 & 0 & 0 \\ -1 & 2 & 0 & 0 & 0 & 0 \\ 0 & 0 & 4 & -2 & 0 & 0 \\ 0 & 0 & -2 & 4 & 0 & 0 \\ 0 & 0 & 0 & 0 & 2 & -1 \\ 0 & 0 & 0 & 0 & -1 & 2 \end{bmatrix} \times \begin{bmatrix} 0 \\ 3 \\ -3 \\ 3 \\ -3 \\ 0 \end{bmatrix}$$

$$[a^{00}] = \frac{72}{EI}$$

$$[a^{00}]^{-1} = \frac{EI}{72}$$

$$[a^{0*}] = [b^0]^T[\alpha][b^*]$$

$$[a^{0*}] = \frac{1}{EI}\begin{bmatrix} 3 & 3 \end{bmatrix}$$

$$\{F^0\} = -[a^{00}]^{-1}[a^{0*}]\{F^*\}$$

$$\{F^0\} = \{-2.5\} = \{F_3^0\}$$

Element Forces

$$\{P\} = [b]\{F\}$$

$$\begin{Bmatrix} P_1 \\ P_2 \\ P_3 \\ P_4 \\ P_5 \\ P_6 \end{Bmatrix} = \begin{bmatrix} 0 & 0 & 0 \\ 1 & 0 & 3 \\ 0 & 0 & -3 \\ 0 & 0 & 3 \\ 0 & -1 & -3 \\ 0 & 0 & 0 \end{bmatrix} \times \begin{Bmatrix} 30 \\ 30 \\ -2.5 \end{Bmatrix} = \begin{Bmatrix} 0 \\ 22.5 \\ 7.5 \\ -7.5 \\ -22.5 \\ 0 \end{Bmatrix}$$

Final Forces

$\{P^f\} = \{P\} - \{P^e\}$

$$= \begin{Bmatrix} 0 \\ 22.5 \\ 7.5 \\ -7.5 \\ -22.5 \\ 0 \end{Bmatrix} - \begin{Bmatrix} 0 \\ 0 \\ 30 \\ -30 \\ 0 \\ 0 \end{Bmatrix} = \begin{Bmatrix} 0 \\ 22.5 \\ -22.5 \\ 22.5 \\ -22.5 \\ 0 \end{Bmatrix}$$

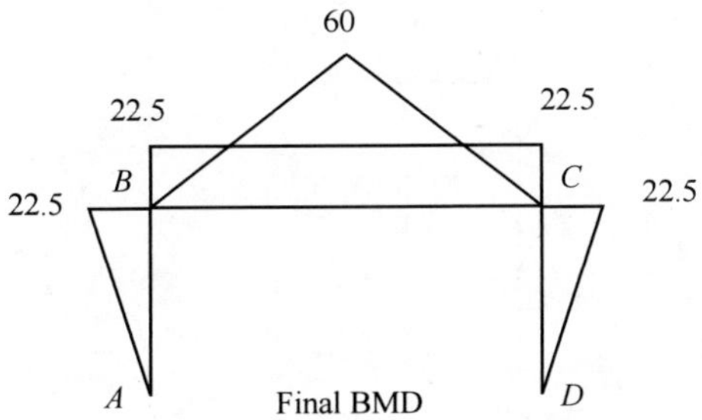

Fig. 2.8(h)

Example 2.7 Analyze the given frame as shown in Fig. 2.9(a) by matrix flexibility method.

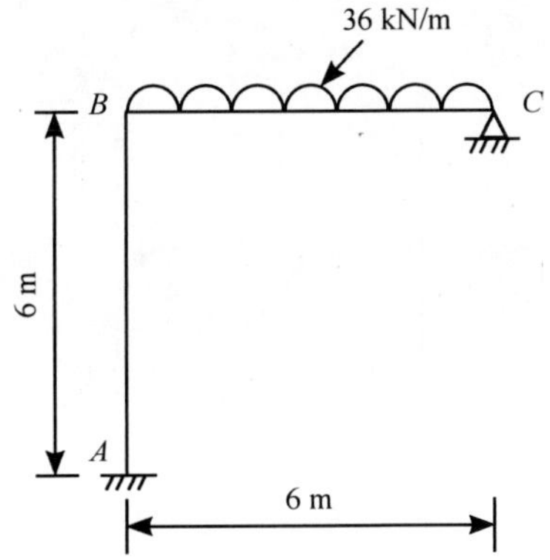

Fig. 2.9(a)

Solution:

Fixed end moments

$$MF_{AB} = MF_{BA} = 0$$

$$MF_{BC} = -\frac{wl^2}{12} = -108 \text{ kN-m}$$

$$MF_{CB} = \frac{wl^2}{12} = 108 \text{ kN-m}$$

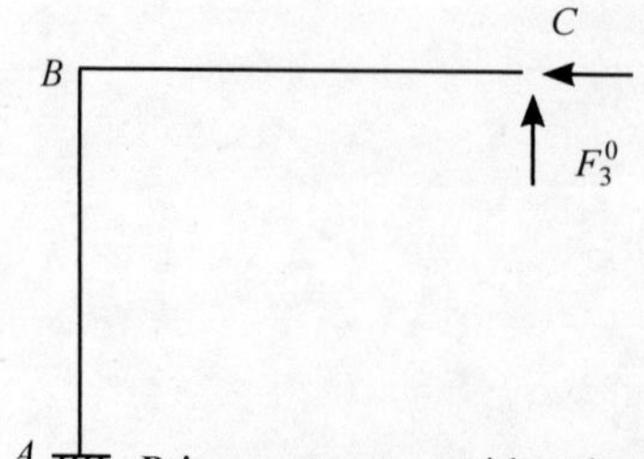

Primary structure with redundants

Fig. 2.9(b)

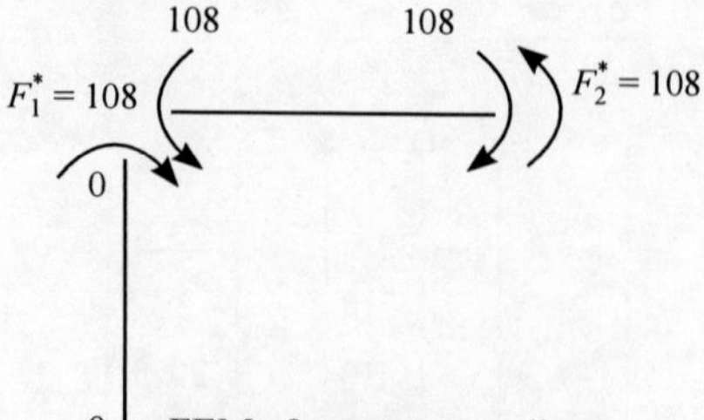

FEMs & system coordinates

Fig. 2.9(c)

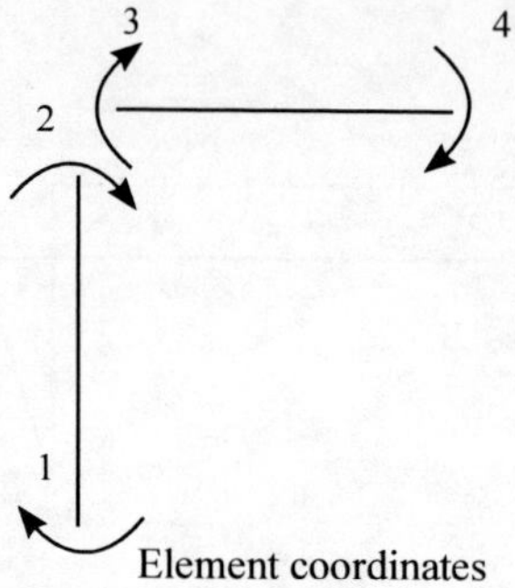

Element coordinates

Fig. 2.9(d)

Formation of [*b*] – Matrix

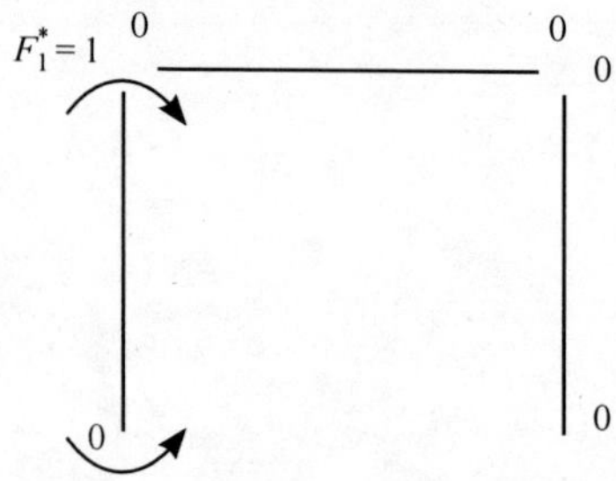

Fig. 2.9(e)

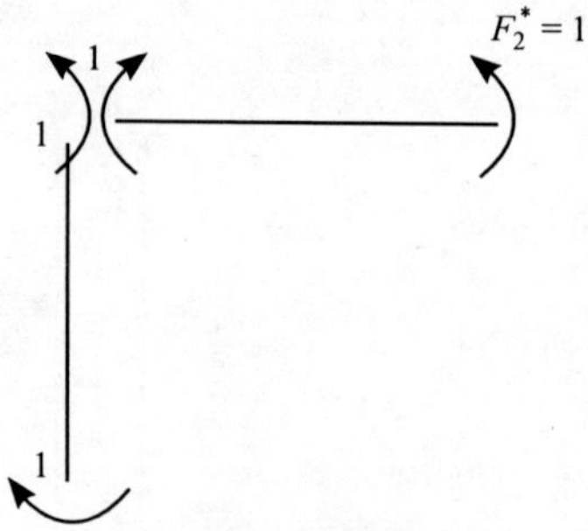

Fig. 2.9(f)

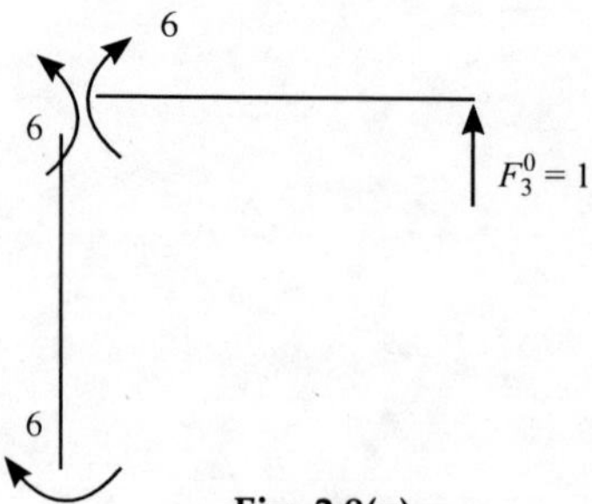

Fig. 2.9(g)

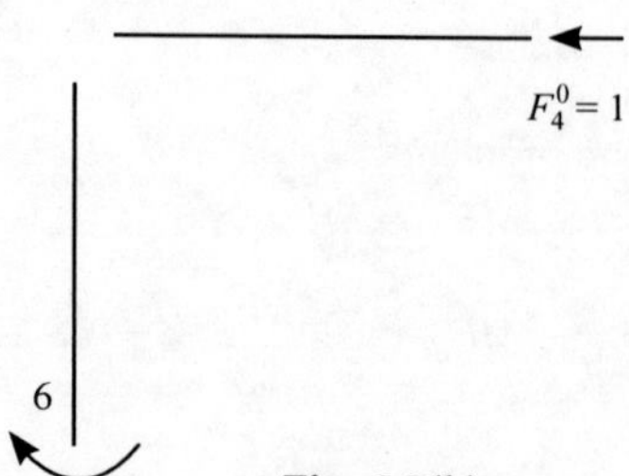

Fig. 2.9(h)

$$[b] = \left[\begin{array}{cc|cc} -1 & 1 & 6 & 6 \\ 1 & -1 & -6 & 0 \\ 0 & 1 & 6 & 0 \\ 0 & -1 & 0 & 0 \end{array}\right]$$
$$\qquad\qquad b^* \qquad\qquad b^0$$

Element Flexibility Matrix

$$[\alpha_1] = \frac{l_1}{6EI}\begin{bmatrix} 2 & -1 \\ -1 & 2 \end{bmatrix} = \frac{1}{EI}\begin{bmatrix} 2 & -1 \\ -1 & 2 \end{bmatrix} = [\alpha_2]$$

Assembled element flexibility matrix

$$[\alpha] = \frac{1}{EI}\begin{bmatrix} 2 & -1 & 0 & 0 \\ -1 & 2 & 0 & 0 \\ 0 & 0 & 2 & -1 \\ 0 & 0 & -1 & 2 \end{bmatrix}$$

Redundant Forces

$$\{F^0\} = -[a^{00}]^{-1}[a^{0*}]\{F^*\}$$

$$\{F^*\} = \begin{Bmatrix} 108 \\ 108 \end{Bmatrix}$$

$$[a^{00}] = [b^0]^T[\alpha][b^0]$$

$$[a^{00}] = \begin{bmatrix} 6 & -6 & 6 & 0 \\ 6 & 0 & 0 & 0 \end{bmatrix} \times \frac{1}{EI}\begin{bmatrix} 2 & -1 & 0 & 0 \\ -1 & 2 & 0 & 0 \\ 0 & 0 & 2 & -1 \\ 0 & 0 & -1 & 2 \end{bmatrix} \times \begin{bmatrix} 6 & 6 \\ -6 & 0 \\ 6 & 0 \\ 0 & 0 \end{bmatrix}$$

$$[a^{00}] = \frac{1}{EI}\begin{bmatrix} 288 & 108 \\ 108 & 72 \end{bmatrix}$$

$$[a^{00}]^{-1} = \frac{EI}{9072}\begin{bmatrix} 72 & -108 \\ -108 & 288 \end{bmatrix}$$

$$[a^{0*}] = [b^0]^T[\alpha][b^*]$$

$$[a^{0*}] = \begin{bmatrix} 6 & -6 & 6 & 0 \\ 6 & 0 & 0 & 0 \end{bmatrix} \times \frac{1}{EI}\begin{bmatrix} 2 & -1 & 0 & 0 \\ -1 & 2 & 0 & 0 \\ 0 & 0 & 2 & -1 \\ 0 & 0 & -1 & 2 \end{bmatrix} \times \begin{bmatrix} -1 & 1 \\ 1 & -1 \\ 0 & 1 \\ 0 & -1 \end{bmatrix}$$

$$[a^{0*}] = \frac{1}{EI}\begin{bmatrix} -36 & 54 \\ -18 & 18 \end{bmatrix}$$

$$\{F^0\} = -[a^{00}]^{-1}[a^{0*}]\{F^*\}$$

$$\{F^0\} = \begin{Bmatrix} -15.428 \\ 23.143 \end{Bmatrix} = \begin{Bmatrix} F_3^0 \\ F_4^0 \end{Bmatrix}$$

Element Forces

$$\{P\} = [b]\{F\}$$

$$\begin{Bmatrix} \\ \\ \\ \\ \end{Bmatrix} = \begin{bmatrix} -1 & 1 & 6 & 6 \\ 1 & -1 & -6 & 0 \\ 0 & 1 & 6 & 0 \\ 0 & -1 & 0 & 0 \end{bmatrix} \times \begin{Bmatrix} 108 \\ 108 \\ -15.428 \\ 23.143 \end{Bmatrix} = \begin{Bmatrix} 46.29 \\ 92.568 \\ 15.432 \\ -108 \end{Bmatrix}$$

Final Forces

$$\{P^f\} = \{P\} - \{P^e\}$$

$$\{P^f\} = \begin{Bmatrix} 46.29 \\ 92.568 \\ 15.432 \\ -108 \end{Bmatrix} - \begin{Bmatrix} 0 \\ 0 \\ 108 \\ -108 \end{Bmatrix} = \begin{Bmatrix} 46.29 \\ 92.568 \\ -92.568 \\ 0 \end{Bmatrix}$$

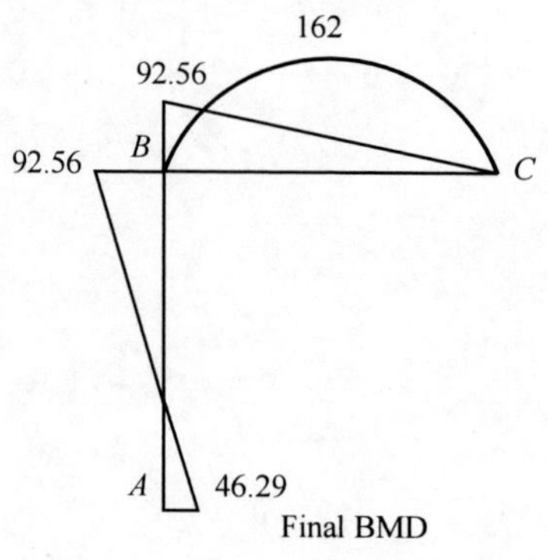

Fig. 2.9(i)

Example 2.8 Analyze the given frame as shown in Fig. 2.10(a) by matrix flexibility method.

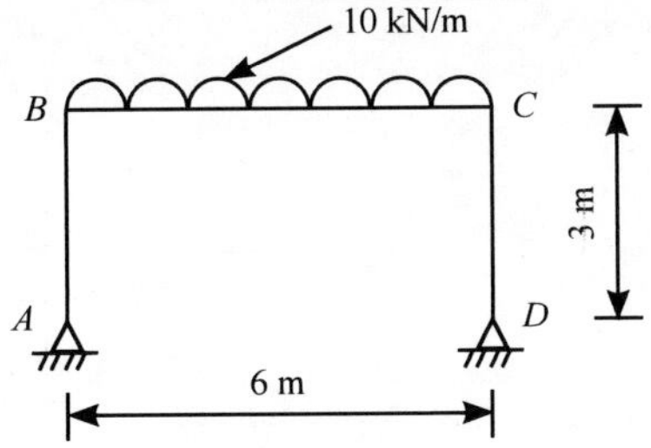

Fig. 2.10(a)

Solution:

Static indeterminacy = 1

Treating the horizontal reaction at D as the redundant, the primary structure is as shown in Fig. 2.10(b).

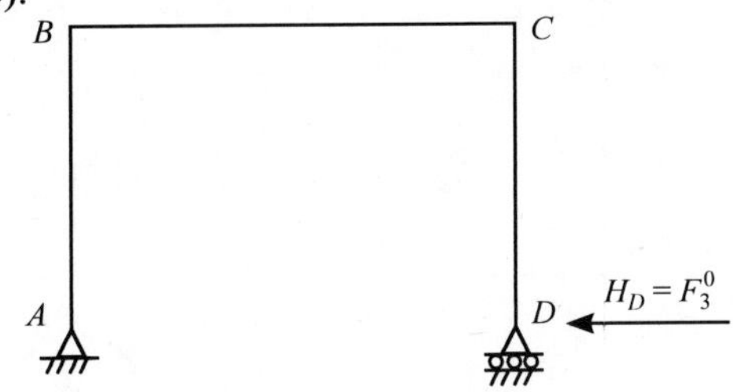

Primary structure with redundant force

Fig. 2.10(b)

Fixed end moments

$$MF_{AB} = MF_{BA} = MF_{CD} = MF_{DC} = 0$$

$$MF_{BC} = -\frac{wl^2}{12} = -30 \text{ kN-m}$$

$$MF_{CB} = \frac{wl^2}{12} = 30 \text{ kN-m}$$

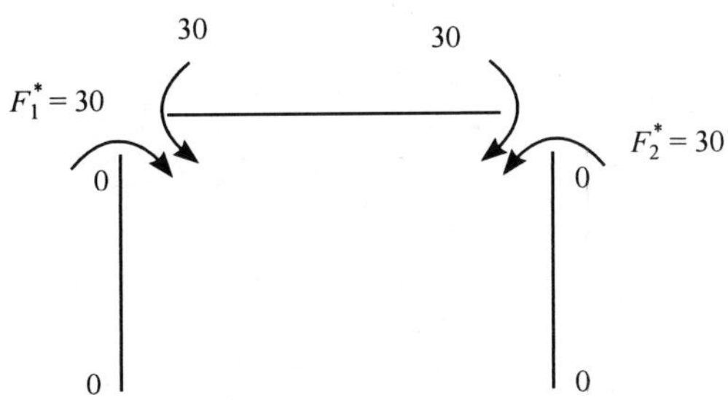

FEMs & system coordinates

Fig. 2.10(c)

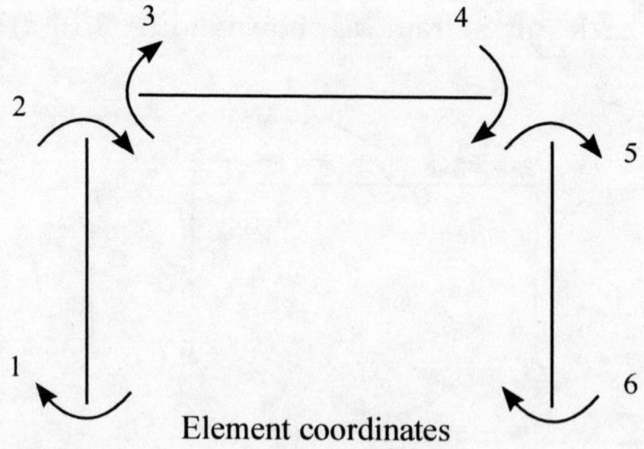

Fig. 2.10(d)

Formation of [*b*] – Matrix

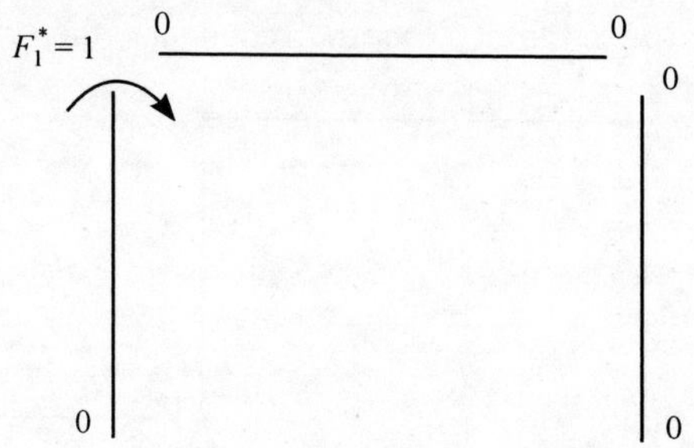

Fig. 2.10(e)

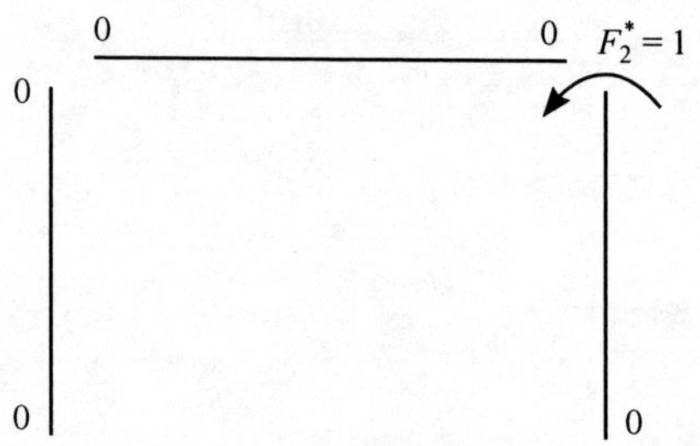

Fig. 2.10(f)

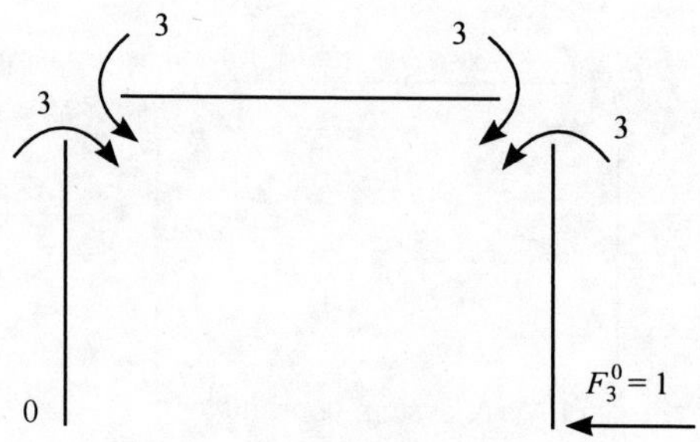

Fig. 2.10(g)

$$[b] = \left[\begin{array}{cc|c} 0 & 0 & 0 \\ 1 & 0 & 3 \\ 0 & 0 & -3 \\ 0 & 0 & 3 \\ 0 & -1 & -3 \\ 0 & 0 & 0 \end{array}\right]$$
$$\qquad\qquad b^* \qquad\qquad b^0$$

Element Flexibility Matrix

$$[\alpha_1] = \frac{l_1}{6E_1I_1}\begin{bmatrix} 2 & -1 \\ -1 & 2 \end{bmatrix} = \frac{1}{EI}\begin{bmatrix} 1 & -0.5 \\ -0.5 & 1 \end{bmatrix} = [\alpha_3]$$

$$[\alpha_2] = \frac{l_2}{6E_2I_2}\begin{bmatrix} 2 & -1 \\ -1 & 2 \end{bmatrix} = \frac{1}{EI}\begin{bmatrix} 2 & -1 \\ -1 & 2 \end{bmatrix}$$

Assembled Element Flexibility Matrix

$$[\alpha] = \frac{1}{EI}\begin{bmatrix} 1 & -0.5 & 0 & 0 & 0 & 0 \\ -0.5 & 1 & 0 & 0 & 0 & 0 \\ 0 & 0 & 2 & -1 & 0 & 0 \\ 0 & 0 & -1 & 2 & 0 & 0 \\ 0 & 0 & 0 & 0 & 1 & -0.5 \\ 0 & 0 & 0 & 0 & -0.5 & 1 \end{bmatrix}$$

Redundant Forces

$$\{F^0\} = -\,[a^{00}]^{-1}[a^{0*}]\{F^*\}$$

$$\{F^*\} = \begin{Bmatrix} 30 \\ 30 \end{Bmatrix}$$

$$[a^{00}] = [b^0]^T[\alpha][b^0]$$

$$[a^{00}] = \begin{bmatrix} 0 & 3 & -3 & 3 & -3 & 0 \end{bmatrix} \times \frac{1}{EI}\begin{bmatrix} 1 & -0.5 & 0 & 0 & 0 & 0 \\ -0.5 & 1 & 0 & 0 & 0 & 0 \\ 0 & 0 & 2 & -1 & 0 & 0 \\ 0 & 0 & -1 & 2 & 0 & 0 \\ 0 & 0 & 0 & 0 & 1 & -0.5 \\ 0 & 0 & 0 & 0 & -0.5 & 1 \end{bmatrix} \times \begin{bmatrix} 0 \\ 3 \\ -3 \\ 3 \\ -3 \\ 0 \end{bmatrix}$$

$$[a^{00}] = \frac{72}{EI}$$

$$[a^{00}]^{-1} = \frac{EI}{72}$$

$$[a^{0*}] = [b^0]^T[\alpha][b^*]$$

$$[a^{0*}] = \begin{bmatrix} 0 & 3 & -3 & 3 & -3 & 0 \end{bmatrix} \times \frac{1}{EI}\begin{bmatrix} 1 & -0.5 & 0 & 0 & 0 & 0 \\ -0.5 & 1 & 0 & 0 & 0 & 0 \\ 0 & 0 & 2 & -1 & 0 & 0 \\ 0 & 0 & -1 & 2 & 0 & 0 \\ 0 & 0 & 0 & 0 & 1 & -0.5 \\ 0 & 0 & 0 & 0 & -0.5 & 1 \end{bmatrix} \times \begin{bmatrix} 0 & 0 \\ 1 & 0 \\ 0 & 0 \\ 0 & 0 \\ 0 & -1 \\ 0 & 0 \end{bmatrix}$$

$$[a^{0*}] = \frac{1}{EI}\begin{bmatrix} 3 & 3 \end{bmatrix}$$

$$\{F^0\} = -[a^{00}]^{-1}[a^{0*}]\{F^*\}$$

$$\{F^*\} = \begin{Bmatrix} 30 \\ 30 \end{Bmatrix}$$

$$\{F^0\} = \{-2.5\} = \{F_3^0\}$$

Element Forces

$$\{P\} = [b]\{F\}$$

$$\begin{Bmatrix} P_1 \\ P_2 \\ P_3 \\ P_4 \\ P_5 \\ P_6 \end{Bmatrix} = \begin{bmatrix} 0 & 0 & 0 \\ 1 & 0 & 3 \\ 0 & 0 & -3 \\ 0 & 0 & 3 \\ 0 & -1 & -3 \\ 0 & 0 & 0 \end{bmatrix} \times \begin{Bmatrix} 30 \\ 30 \\ -2.5 \end{Bmatrix} = \begin{Bmatrix} 0 \\ 22.5 \\ 7.5 \\ -7.5 \\ -22.5 \\ 0 \end{Bmatrix}$$

Final Forces

$$\{P^f\} = \{P\} - \{P^e\}$$

$$= \begin{Bmatrix} 0 \\ 22.5 \\ 7.5 \\ -7.5 \\ -22.5 \\ 0 \end{Bmatrix} - \begin{Bmatrix} 0 \\ 0 \\ 30 \\ -30 \\ 0 \\ 0 \end{Bmatrix} = \begin{Bmatrix} \\ 22.5 \\ 22.5 \\ 22.5 \\ 22.5 \\ \end{Bmatrix}$$

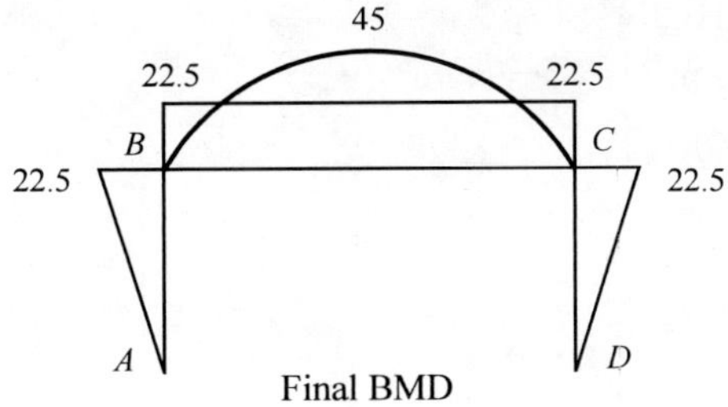

Final BMD

Fig. 2.10(h)

2.4 ANALYSIS OF PIN-JOINTED PLANE FRAMES

Pin-jointed plane frames are analysed here by flexibility matrix method.

Example 2.9 Find the forces in all the members of the pin-jointed frame as shown in Fig. 2.11(a) using flexibility method.

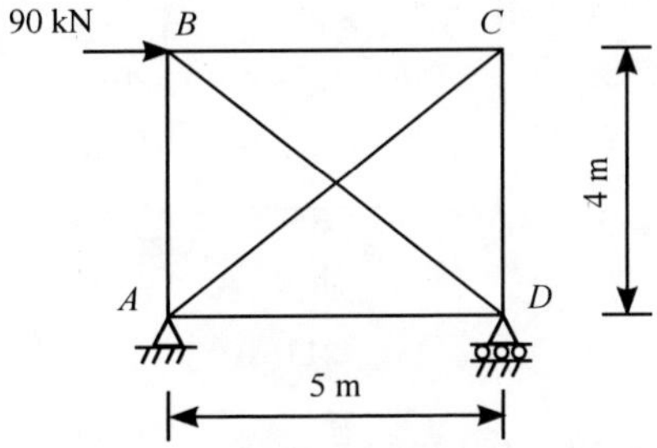

Fig. 2.11(a)

Solution:

Internal indeterminacy = 1

External indeterminacy = 0

Treating member *AC* as redundant and removing it, the statically determinate structure is as shown in Fig. 2.11(b).

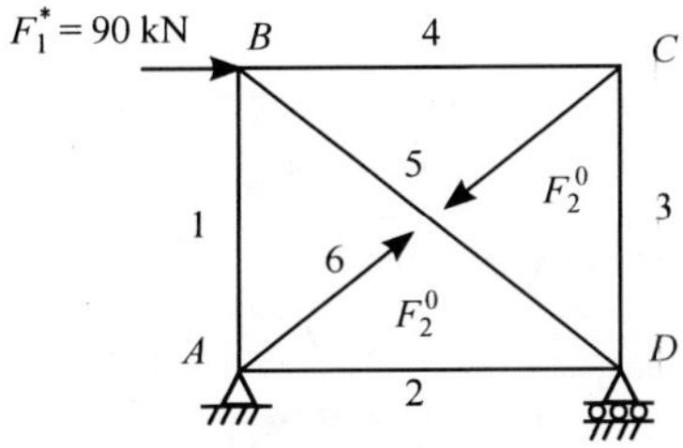

Primary structure and $\{F\}$ system

Fig. 2.11(b)

Transformation Matrix [*b*]

Applying $F_1^* = 1$ in the primary structure

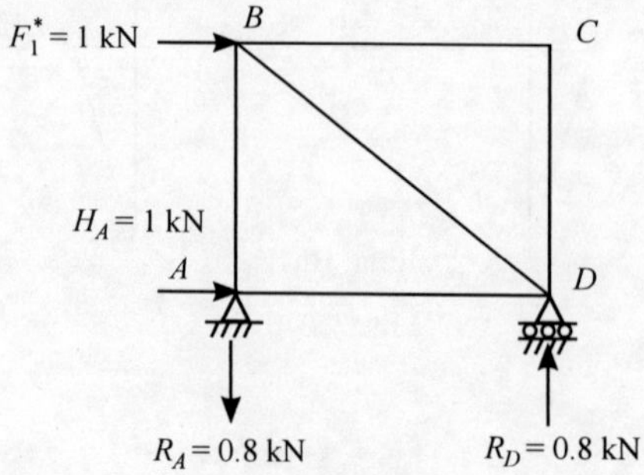

Fig. 2.11(c)

b^* - values are got by analysing Fig. 2.11(c) by using method of joints.

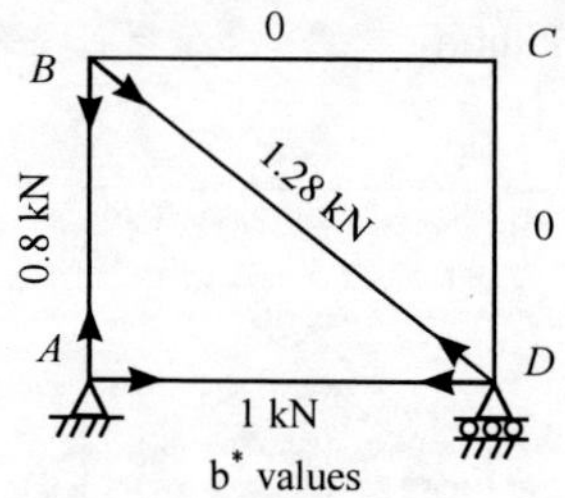

Fig. 2.11(d)

Applying $F_2^0 = 1$ in the primary structure

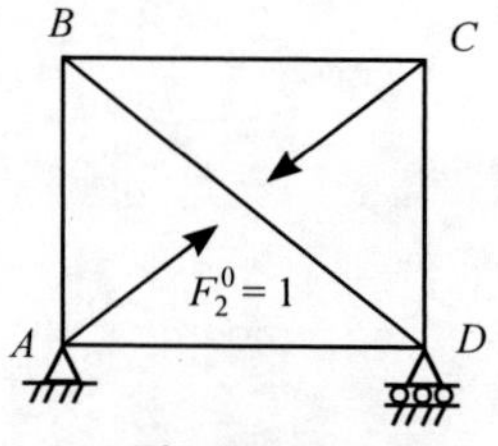

Fig. 2.11(e)

b^0 - values are got by analysing the Fig. 2.11(e) by using method of joints.

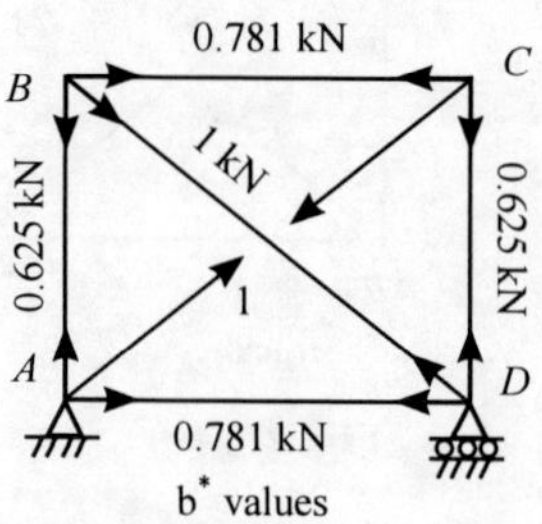

Fig. 2.11(f)

$$[b] = \left[\begin{array}{c|c} 0.8 & -0.625 \\ 1 & -0.781 \\ 0 & -0.625 \\ 0 & -0.781 \\ -1.28 & 1 \\ 0 & 1 \end{array}\right]$$
$$\begin{array}{cc} b^* & b^0 \end{array}$$

Assembled Element Flexibility Matrix [α]

$$[\alpha] = \frac{1}{AE}\begin{bmatrix} 4 & 0 & 0 & 0 & 0 & 0 \\ 0 & 5 & 0 & 0 & 0 & 0 \\ 0 & 0 & 4 & 0 & 0 & 0 \\ 0 & 0 & 0 & 5 & 0 & 0 \\ 0 & 0 & 0 & 0 & 6.4 & 0 \\ 0 & 0 & 0 & 0 & 0 & 6.4 \end{bmatrix}$$

Redundant Forces

$$\{F^0\} = -[a^{00}]^{-1}[a^{0*}]\{F^*\}$$

$$[a^{00}] = [b^0]^T[\alpha][b^0]$$

$$[a^{00}] = \begin{bmatrix} -0.625 & -0.781 & -0.625 & -0.781 & 1 & 1 \end{bmatrix} \times \frac{1}{AE}\begin{bmatrix} 4 & 0 & 0 & 0 & 0 & 0 \\ 0 & 5 & 0 & 0 & 0 & 0 \\ 0 & 0 & 4 & 0 & 0 & 0 \\ 0 & 0 & 0 & 5 & 0 & 0 \\ 0 & 0 & 0 & 0 & 6.4 & 0 \\ 0 & 0 & 0 & 0 & 0 & 6.4 \end{bmatrix} \times \begin{bmatrix} -0.625 \\ -0.781 \\ -0.625 \\ -0.781 \\ 1 \\ 1 \end{bmatrix}$$

$$[a^{00}] = \frac{22.03}{AE}$$

$$[a^{00}]^{-1} = \frac{AE}{22.03}$$

$$[a^{0*}] = [b^0]^T[\alpha][b^*]$$

$$[a^{0*}] = \begin{bmatrix} -0.625 & -0.781 & -0.625 & -0.781 & 1 & 1 \end{bmatrix} \times \frac{1}{AE}\begin{bmatrix} 4 & 0 & 0 & 0 & 0 & 0 \\ 0 & 5 & 0 & 0 & 0 & 0 \\ 0 & 0 & 4 & 0 & 0 & 0 \\ 0 & 0 & 0 & 5 & 0 & 0 \\ 0 & 0 & 0 & 0 & 6.4 & 0 \\ 0 & 0 & 0 & 0 & 0 & 6.4 \end{bmatrix} \times \begin{bmatrix} 0.8 \\ 1 \\ 0 \\ 0 \\ -1.28 \\ 0 \end{bmatrix}$$

$$[a^{0*}] = -\frac{14.10}{AE}$$

$$\{F^0\} = -[a^{00}]^{-1}[a^{0*}]\{F^*\}$$

$$\{F^0\} = -\frac{AE}{22.03} \times -\frac{14.10}{AE} \times 90$$

$$\{F^0\} = 57.603 \text{ kN}$$

Element Forces

$$\{P\} = [b]\{F\}$$

$$\begin{Bmatrix} P_1 \\ P_2 \\ P_3 \\ P_4 \\ P_5 \\ P_6 \end{Bmatrix} = \begin{bmatrix} 0.8 & -0.625 \\ 1 & -0.781 \\ 0 & -0.625 \\ 0 & -0.781 \\ -1.28 & 1 \\ 0 & 1 \end{bmatrix} \times \begin{Bmatrix} 90 \\ 57.603 \end{Bmatrix} = \begin{Bmatrix} 36 \\ 45 \\ -36 \\ -45 \\ -57.6 \\ 57.6 \end{Bmatrix}$$

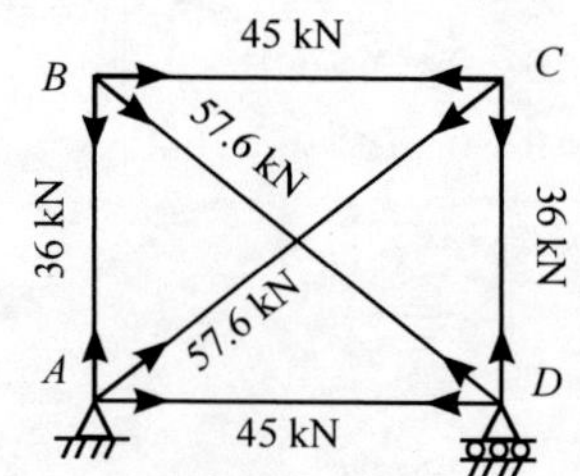

Final values

Fig. 2.11(g)

Example 2.10 Find the forces in all the members of the pin-jointed frame as shown in Fig. 2.12(a) using flexibility method.

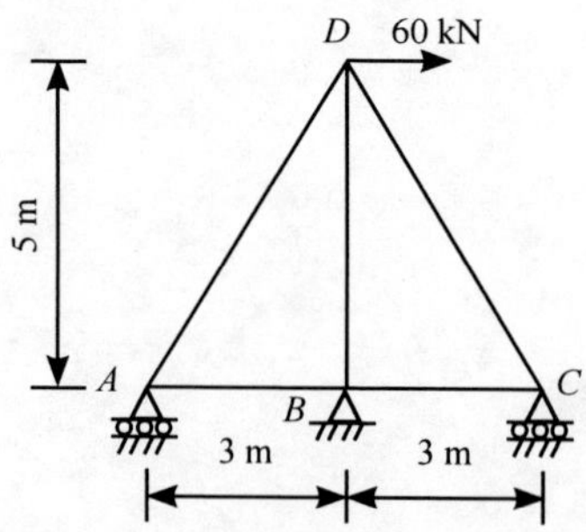

Fig. 2.12(a)

Solution:

Internal indeterminacy = 0

External indeterminacy = 1

The structure is externally indeterminate to degree one. Treating support at A as redundant, the primary structure is as shown in Fig. 2.12(b).

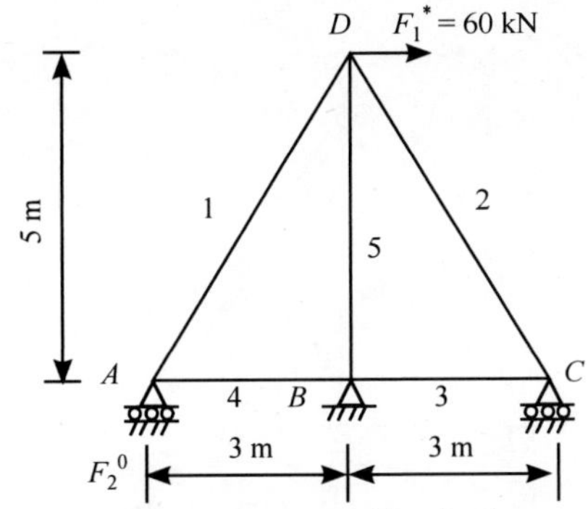

Primary structure with redundants

Fig. 2.12(b)

Transformation Matrix [*b*]

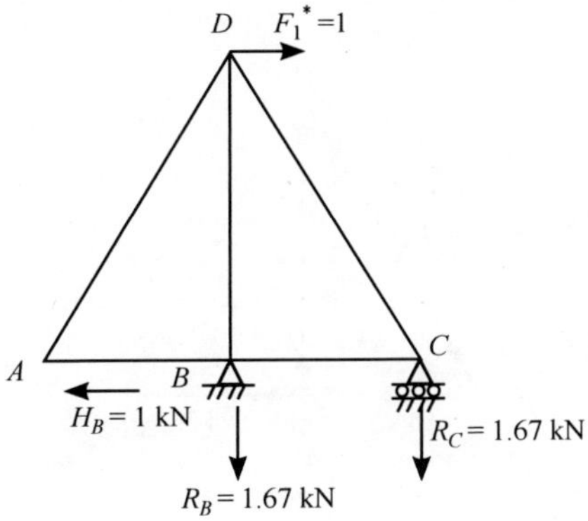

Fig. 2.12(c)

b^* - values are got by analysing the Fig. 2.12(c) by using method of joints.

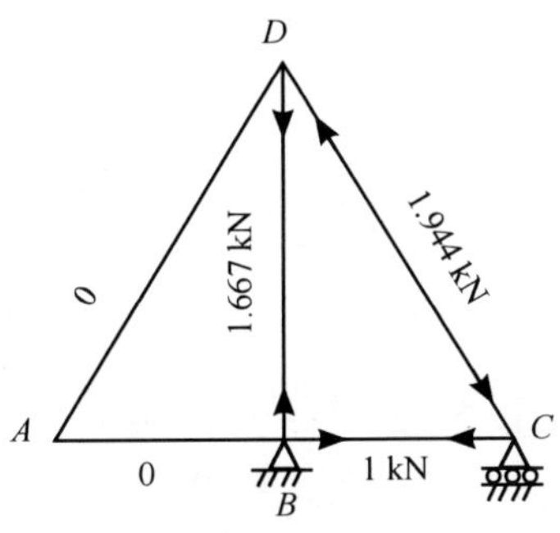

b^* values

Fig. 2.12(d)

Apply $F_2^0 = 1$

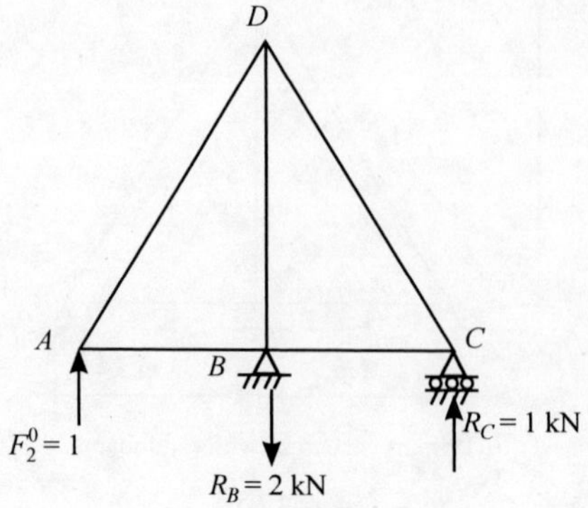

Fig. 2.12(e)

b^0 - values are got by analysing Fig. 2.12(e) by using method of joints.

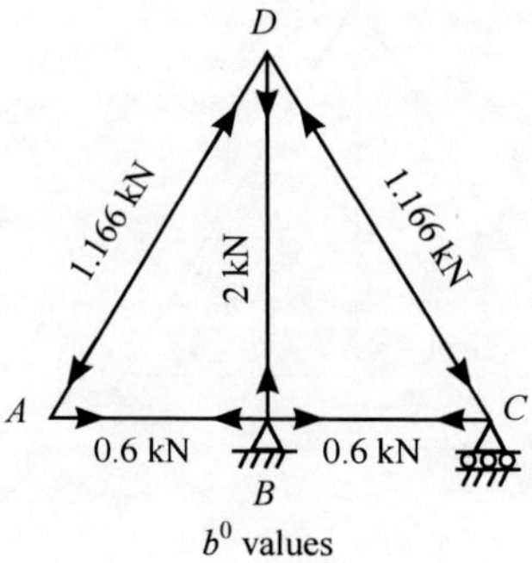

Fig. 2.12(f)

$$[b] = \left[\begin{array}{c:c} 0 & -1.166 \\ -1.944 & -1.166 \\ 1 & 0.6 \\ 0 & 0.6 \\ 1.66 & 2 \end{array}\right]$$
$$\quad b^* \qquad b^0$$

Assembled Element Flexibility Matrix [α]

$$[\alpha] = \frac{1}{AE}\begin{bmatrix} 5.83 & 0 & 0 & 0 & 0 \\ 0 & 5.83 & 0 & 0 & 0 \\ 0 & 0 & 3 & 0 & 0 \\ 0 & 0 & 0 & 3 & 0 \\ 0 & 0 & 0 & 0 & 5 \end{bmatrix}$$

Redundant Forces

$$\{F^0\} = -[a^{00}]^{-1}[a^{0*}]\{F^*\}$$

$$[a^{00}] = [b^0]^T[\alpha][b^0]$$

$$[a^{00}] = \begin{bmatrix} -1.166 & -1.166 & 0.6 & 0.6 & 2 \end{bmatrix} \times \frac{1}{AE} \begin{bmatrix} 5.83 & 0 & 0 & 0 & 0 \\ 0 & 5.83 & 0 & 0 & 0 \\ 0 & 0 & 3 & 0 & 0 \\ 0 & 0 & 0 & 3 & 0 \\ 0 & 0 & 0 & 0 & 5 \end{bmatrix} \times \begin{bmatrix} -1.166 \\ -1.166 \\ 0.6 \\ 0.6 \\ 2 \end{bmatrix}$$

$$[a^{00}] = \frac{38.01}{AE}$$

$$[a^{00}]^{-1} = \frac{AE}{38.01}$$

$$[a^{0*}] = [b^0]^T[\alpha][b^*]$$

$$[a^{0*}] = \begin{bmatrix} -1.166 & -1.166 & 0.6 & 0.6 & 2 \end{bmatrix} \times \frac{1}{AE} \begin{bmatrix} 5.83 & 0 & 0 & 0 & 0 \\ 0 & 5.83 & 0 & 0 & 0 \\ 0 & 0 & 3 & 0 & 0 \\ 0 & 0 & 0 & 3 & 0 \\ 0 & 0 & 0 & 0 & 5 \end{bmatrix} \times \begin{bmatrix} 0 \\ -1.944 \\ 1 \\ 0 \\ 1.66 \end{bmatrix}$$

$$[a^{0*}] = \frac{31.59}{AE}$$

$$\{F^0\} = -[a^{00}]^{-1}[a^{0*}]\{F^*\}$$

$$\{F^0\} = -\frac{AE}{38.01} \times \frac{31.59}{AE} \times 60$$

$$\{F^0\} = -49.87 \text{ kN}$$

Element Forces

$$\{P\} = [b]\{F\}$$

$$\{P\} = \begin{bmatrix} 0 & -1.166 \\ -1.944 & -1.166 \\ 1 & 0.6 \\ 0 & 0.6 \\ 1.66 & 2 \end{bmatrix} \times \begin{Bmatrix} 60 \\ -49.87 \end{Bmatrix} = \begin{Bmatrix} 58.15 \\ -58.25 \\ 30.08 \\ -29.92 \\ -0.26 \end{Bmatrix}$$

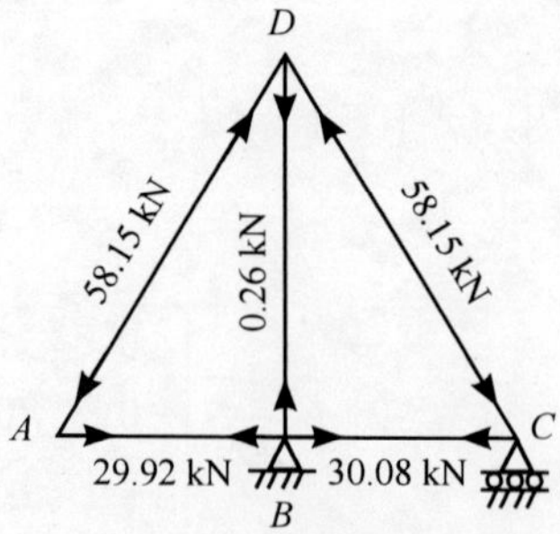

Final values

Fig. 2.12(g)

2.5 BRIEF PROCEDURE FOR ANALYSIS OF INDETERMINATE STRUCTURES

In general, the following procedure is to be followed.

1. Identify the static indeterminacy, decide the primary structure and indicate the redundants $\{F^0\}$.
2. Select $\{P\}$ coordinates for the internal forces.
3. Compile the external forces $\{F^*\}$. $\{F^*\}$ is called system coordinate.

   Now we have a determinate structure with two classes of loads, namely, $\{F^*\}$ and $\{F^0\}$.
4. Generate the $[b]$ matrix such that $[P] = [b]\{F\}$ where,

   $[b]$ = transformation matrix

   $[P]$ = element forces

   $\{F\}$ = system forces
5. Compute system flexibility matrix,

   $$[a] = [b]^T[\alpha][b]$$
   $$[a^{00}] = [b^0]^T[\alpha][b^0]$$
   $$[a^{0*}] = [b^0]^T[\alpha][b^*]$$

   The part of the $[b]$ matrix relating to $\{F^*\}$ is called $[b^*]$ and part of the $[b]$ matrix relating to $\{F^0\}$is called $[b^0]$.
6. Compute redundant forces, $\{F^0\} = -[a^{00}]^{-1}[a^{0*}]\{F^*\}$
7. Compute element forces, i.e., forces along element coordinates, $\{P\} = [b]\{F\}$.
8. Compute final forces, $\{P^f\} = \{P\} - \{P^e\}$ where, $\{P^e\}$ is vector of equivalent joint forces.
9. Finally sketch bending moment diagram.

Questions

1. Derive the flexibility equation.
2. Give step-by-step procedure of analyzing a structure by flexibility matrix method.
3. Analyse the continuous beam shown in Fig. 2.13, by flexibility matrix method. Flexural rigidity EI is constant throughout.

   [Ans: $M_{AB} = -66.67$ kN-m, $M_{BA} = 58.67$ kN-m]

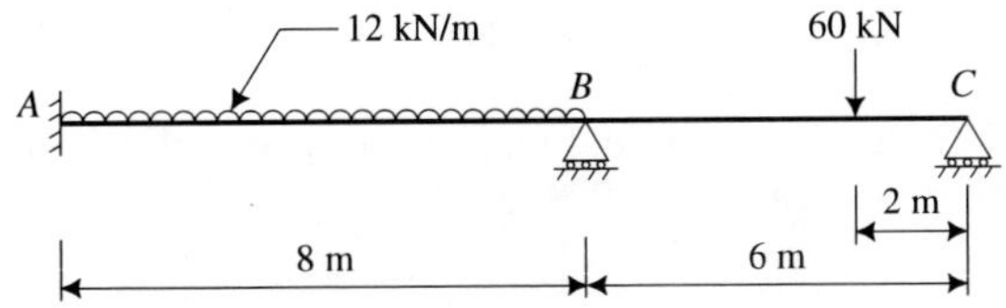

Fig. 2.13

4. Analyse the continuous beam shown in Fig. 2.14 by flexibility method.

$$M_{AB} = -10.83 \text{ kN-m}$$
$$M_{BA} = 62.92 \text{ kN-m}$$
$$M_{CB} = 10.00 \text{ kN-m}$$

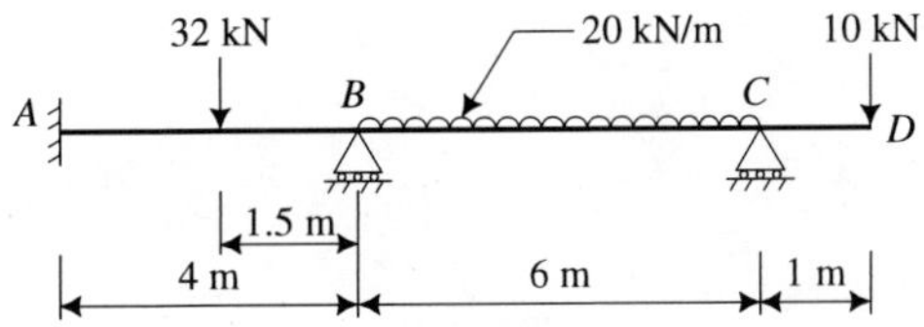

Fig. 2.14

5. Analyse the continuous beam shown in Fig. 2.15 by flexibility matrix method. EI is constant.

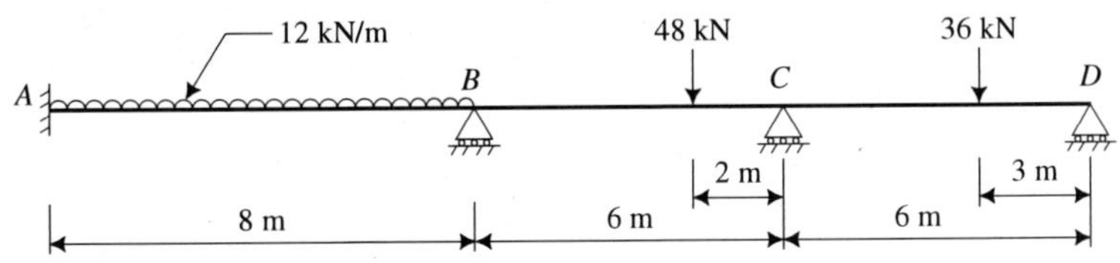

Fig. 2.15

6. Determine the reactions at support D developed due to loading as shown in Fig. 2.16.

$$H_D = 28.55 \text{ kN}$$
$$V_D = 46.33 \text{ kN}$$

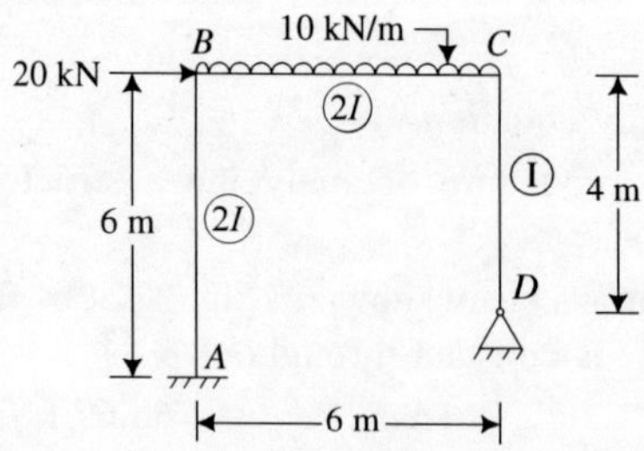

Fig. 2.16

7. Analyse the frame shown in Fig. 2.17 by flexibility matrix method and determine the moments at *A*, *B*, *C* and *D*.

$M_{AB} = -32.14$ kN-m

$M_{BA} = -M_{BC} = -7.14$ kN-m

$M_{CB} = -M_{CD} = 92.86$ kN-m

$M_{DC} = -82.14$ kN

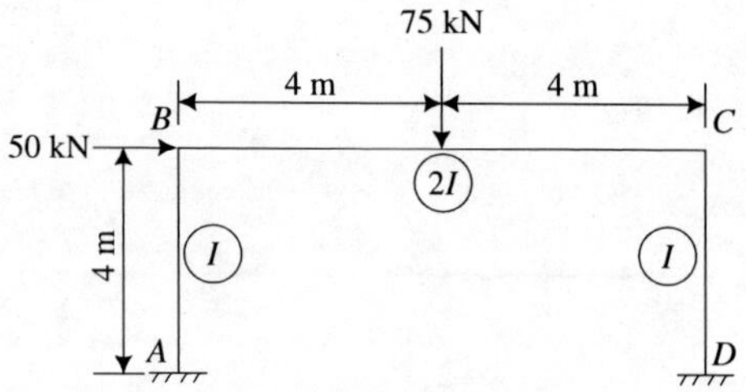

Fig. 2.17

8. Find the forces in all members of the truss shown in Fig. 2.18. Given: Cross-sectional area and modulus of elasticity is same for all members.

Ans. $F_{AB} = -6.85$ kN $\quad F_{AD} = 13.16$ kN

$F_{BC} = 3.16$ kN $\quad F_{BD} = -18.60$ kN

$F_{CD} = 3.64$ kN $\quad F_{AC} = 5.15$ kN

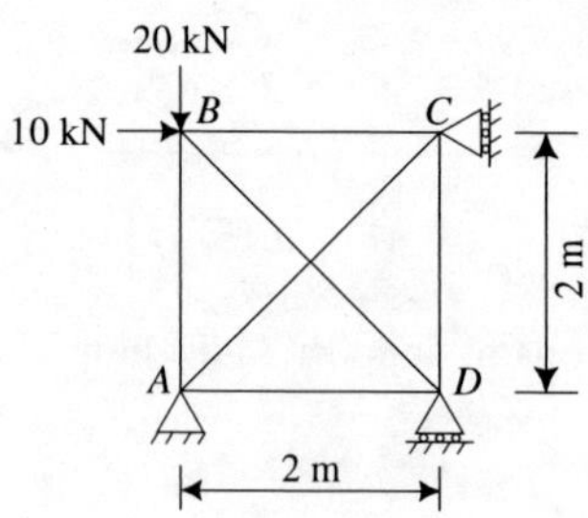

Fig. 2.18

3 Stiffness Matrix Method: Element Approach

The behaviour of the structure can largely be defined by force-displacement relation in the form of a matrix. The systematic development of slope deflection method in the matrix form has resulted in stiffness matrix method. Displacements at the discrete points are taken as the unknowns throughout the structure, equilibrium conditions are enforced resulting in a set of simultaneous equations in which the coefficients are stiffness coefficients. These equations developed in the matrix form are solved for the unknown displacement components from which the force component can be evaluated. This method is known as:

1. Displacement method since displacements are unknown.
2. Equilibrium method since equilibrium conditions are employed.
3. Stiffness method since stiffness coefficients are used.

The element approach which is considered in this chapter is suitable for developing general-purpose computer program. In this chapter, stiffness matrix method by element approach is explained and is used to solve the following types of one-dimensional elements with which readers are familiar to solve:

(a) Beams
(b) Rigid jointed plane frames
(c) Pin-jointed plane frames

3.1 STIFFNESS EQUATION

Stiffness is defined as force per unit displacement.

Stiffness Matrix: Equation used: $[k]\{\Delta\} = \{F\}$

where,

$$\{F\} = \text{Force vector}$$

$$\{F\} = \begin{Bmatrix} F_1 \\ F_2 \\ F_3 \\ \cdot \\ \cdot \\ F_n \end{Bmatrix}$$

These are generalized forces. They may be axial, translation, moment or twist.

$\{\Delta\}$ = Displacement vector

$$\{\Delta\} = \begin{Bmatrix} \Delta_1 \\ \Delta_2 \\ \Delta_3 \\ . \\ . \\ \Delta_n \end{Bmatrix}$$

These are generalized displacements. They may be axial, translation, moment or twist.

$[k]$ = Stiffness matrix

$$[k] = \begin{bmatrix} k_{11} & k_{12} & . & . & k_{1n} \\ k_{21} & k_{22} & . & . & k_{2n} \\ . & . & k_{ij} & . & . \\ k_{n1} & k_{n2} & . & . & k_{nn} \end{bmatrix}$$

where, n = no. of coordinates

k_{ij} = stiffness coefficient

A stiffness coefficient is defined as a coefficient which gives the force at coordinate i due to unit displacement at coordinate j with the condition that no displacement occurs expect at j.

Example 3.1

Develop the stiffness matrix for the beam as shown in Fig. 3.1(a) having constant EI.

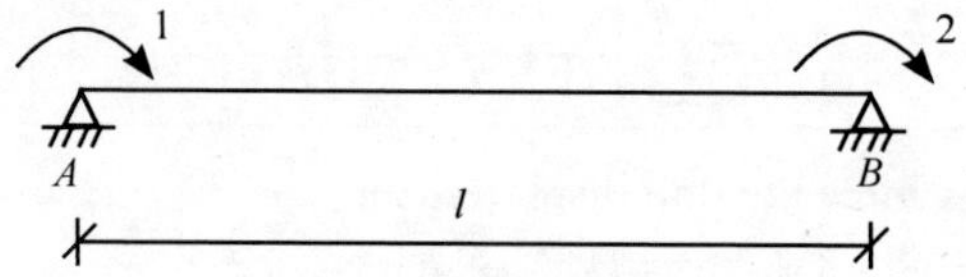

Fig. 3.1(a)

Solution:

$$[k] = \begin{bmatrix} k_{11} & k_{12} \\ k_{21} & k_{22} \end{bmatrix}$$

To develop the first column of the stiffness matrix, first apply the unit displacement along coordinate 1 only with the condition that no displacement occurs at coordinate 2. Refer Fig. 3.1(b).

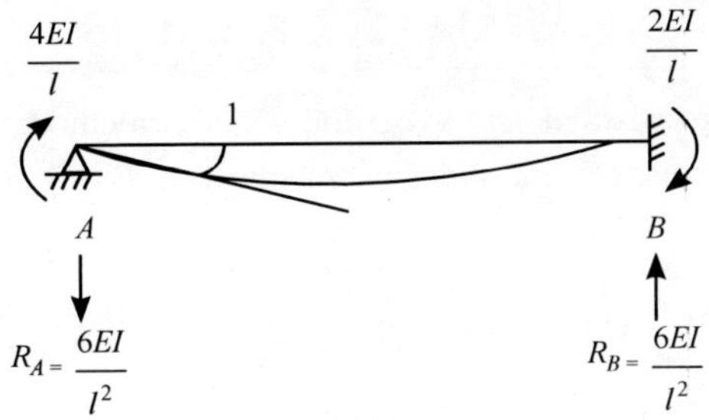

Fig. 3.1(b)

$$k_{11} = \frac{4EI}{l}$$

$$k_{21} = \frac{2EI}{l}$$

Next, to develop the second column of the stiffness matrix, apply unit displacement along coordinate 2 only, with the condition that no displacement occurs at coordinate 1. Refer Fig. 3.1(c).

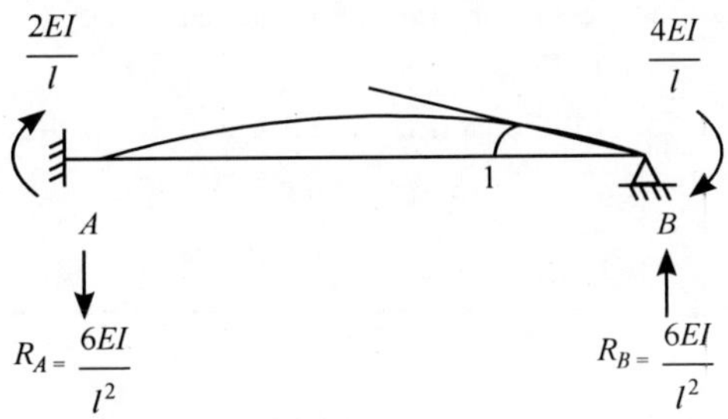

Fig. 3.1(c)

$$k_{12} = \frac{2EI}{l}$$

$$k_{22} = \frac{4EI}{l}$$

$$[k] = \begin{bmatrix} \frac{4EI}{l} & \frac{2EI}{l} \\ \frac{2EI}{l} & \frac{4EI}{l} \end{bmatrix}$$

$$[k] = \frac{EI}{l}\begin{bmatrix} 4 & 2 \\ 2 & 4 \end{bmatrix}$$

3.2 ANALYSIS OF BEAMS

Continuous beams are analysed here by stiffness matrix method.

Example 3.2 Analyze the continuous beam shown in Fig. 3.2(a) flexibility matrix method.

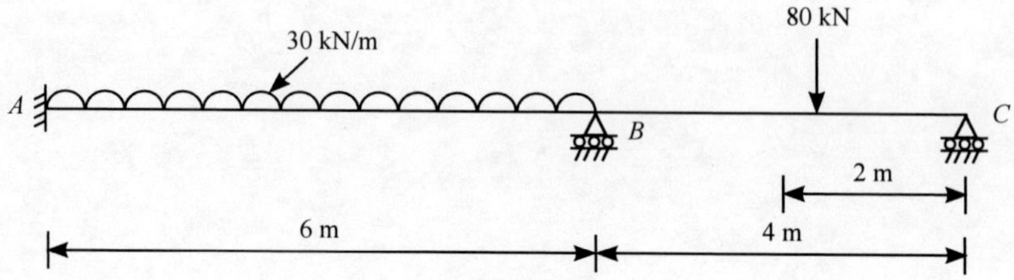

Fig. 3.2(a)

Solution:

Step 1:

Determine the degree of kinematic indeterminacy of the structural system. Identify the independent displacement components. Assign coordinates 1, 2, ..., *n* to the independent displacement components.

For the given problem, the degree of kinematic indeterminacy = 2

Kinematic indeterminacy, $KI = 2$

θ_B and θ_C are the unknown displacements. This will form the system coordinates. Refer Fig. 3.2(b).

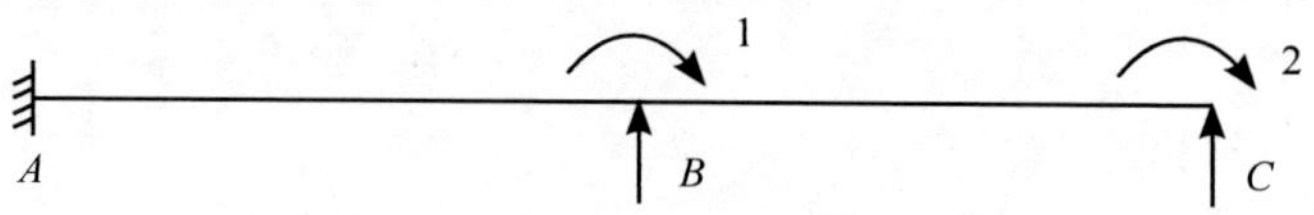

System coordinates

Fig. 3.2(b)

Step 2:

Identify the elements of the structural system and assign the element coordinates or element displacements.

We have to divide the structure into number of elements. Mark all the coordinates in clockwise direction.

For the given problem, the structure has been divided into two elements and we have chosen four element coordinates or element displacements. Refer Fig. 3.2(c).

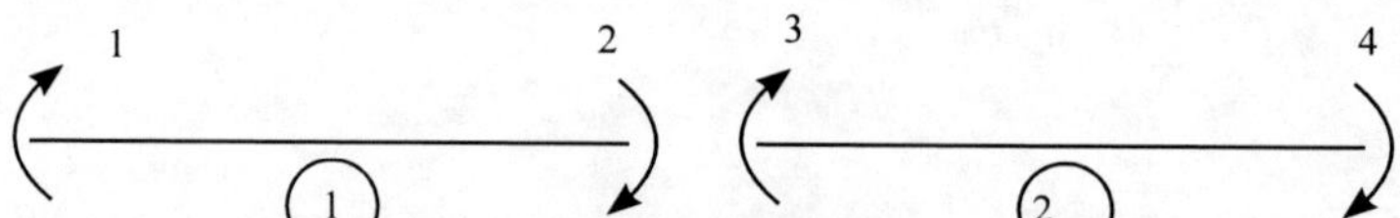

Element coordinates

Fig. 3.2(c)

Step 3: Determine the forces at element coordinates in the fixed state. Note them as $\{p^0\}$, that is, these are fixed end moments.

Following are the fixed end moments for the given problem.

$$MF_{AB} = -\frac{wl^2}{12} = -90 \text{ kN-m}$$

$$MF_{BA} = \frac{wl^2}{12} = 90 \text{ kN-m}$$

$$MF_{BC} = -\frac{wl}{8} = -40 \text{ kN-m}$$

$$MF_{CB} = \frac{wl}{8} = 40 \text{ kN-m}$$

Forces in the element coordinates

$$\{P^0\} = \begin{Bmatrix} -90 \\ 90 \\ -40 \\ 40 \end{Bmatrix}$$

Step 4:

Develop the transformation matrix [β]

Transformation matrix is defined as the matrix connecting vectors in local coordinate system and global coordinate system.

Apply unit displacement along system coordinates and get the displacements along element coordinates.

For the given problem, to get the first column of the [β] matrix, apply an unit displacement at coordinate 1 and get the corresponding element coordinate displacements. Refer Fig. 3.2(d).

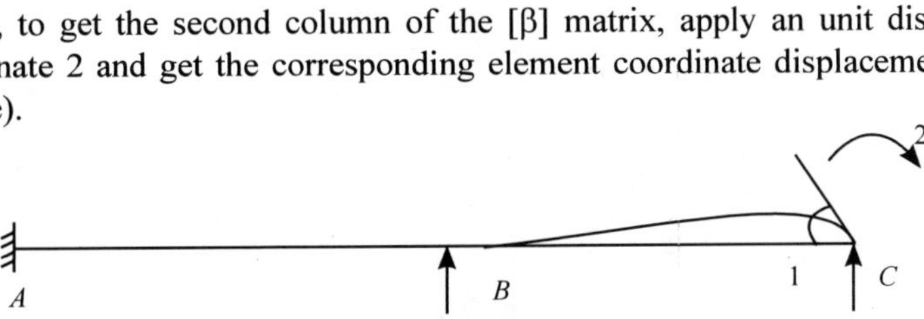

Fig. 3.2(d)

Similarly, to get the second column of the [β] matrix, apply an unit displacement at coordinate 2 and get the corresponding element coordinate displacements. Refer Fig. 3.2(e).

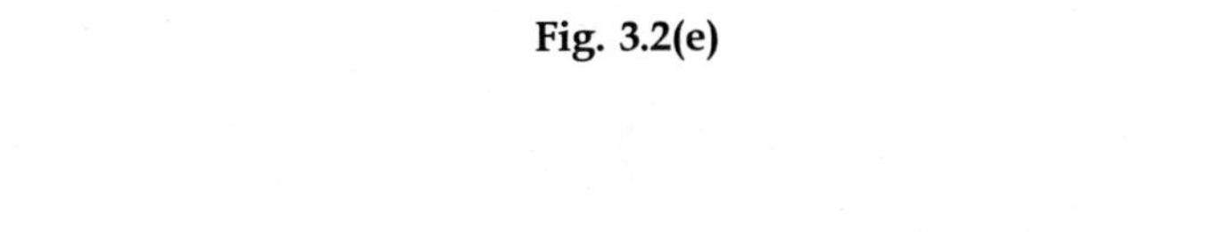

Fig. 3.2(e)

Therefore, $$[\beta] = \begin{bmatrix} 0 & 0 \\ 1 & 0 \\ 1 & 0 \\ 0 & 1 \end{bmatrix}$$

Step 5:

Let the forces in the system coordinates be F^0

$$\{F^0\} = [\beta]^T\{P^0\}$$

$$\{F^0\} = \begin{bmatrix} 0 & 1 & 1 & 0 \\ 0 & 0 & 0 & 1 \end{bmatrix} \times \begin{Bmatrix} -90 \\ 90 \\ -40 \\ 40 \end{Bmatrix} = \begin{Bmatrix} 50 \\ 40 \end{Bmatrix}$$

Let the forces in the system coordinates due to loading be F'

In this case

$\{F'\} = \begin{Bmatrix} 0 \\ 0 \end{Bmatrix}$ (because no forces in the given loading are acting at system coordinates.)

Therefore, system forces,

$$\{F\} = [\{F'\} - \{F^0\}]$$

$$\{F\} = [\begin{Bmatrix} 0 \\ 0 \end{Bmatrix} - \begin{Bmatrix} 50 \\ 40 \end{Bmatrix}] = \begin{Bmatrix} -50 \\ -40 \end{Bmatrix}$$

Step 6:

Form the element stiffness matrix, $[k]$

In the given problem,

For element 1,

$$[k_1] = \frac{E_1 I_1}{l_1}\begin{bmatrix} 4 & 2 \\ 2 & 4 \end{bmatrix} = \frac{EI}{6}\begin{bmatrix} 4 & 2 \\ 2 & 4 \end{bmatrix}$$

For element 2,

$$[k_2] = \frac{E_2 I_2}{l_2}\begin{bmatrix} 4 & 2 \\ 2 & 4 \end{bmatrix} = \frac{EI}{4}\begin{bmatrix} 4 & 2 \\ 2 & 4 \end{bmatrix}$$

Assembled element stiffness matrix, $[k]$

$$[k] = \begin{bmatrix} [k_1] & 0 \\ 0 & [k_2] \end{bmatrix}$$

$$[k] = EI\begin{bmatrix} 0.67 & 0.33 & 0 & 0 \\ 0.33 & 0.67 & 0 & 0 \\ 0 & 0 & 1 & 0.5 \\ 0 & 0 & 0.5 & 1 \end{bmatrix}$$

Step 7:

Form the system stiffness matrix $[k]$ using the relation,

$$[k] = [\beta]^T[k][\beta]$$

$$[k] = \begin{bmatrix} 0 & 1 & 1 & 0 \\ 0 & 0 & 0 & 1 \end{bmatrix} \times EI \begin{bmatrix} 0.67 & 0.33 & 0 & 0 \\ 0.33 & 0.67 & 0 & 0 \\ 0 & 0 & 1 & 0.5 \\ 0 & 0 & 0.5 & 1 \end{bmatrix} \times \begin{bmatrix} 0 & 0 \\ 1 & 0 \\ 1 & 0 \\ 0 & 1 \end{bmatrix}$$

$$[k] = EI\begin{bmatrix} 1.67 & 0.5 \\ 0.5 & 1 \end{bmatrix}$$

Also,

$$[k]^{-1} = \frac{1}{1.42EI}\begin{bmatrix} 1 & -0.5 \\ -0.5 & 1.67 \end{bmatrix}$$

Step 8:

Compute the system displacement vector, $\{\Delta\}$ using the relation,

$$\{\Delta\} = [k]^{-1}[\{F'\}-\{F^0\}]$$

$$\{\Delta\} = [k]^{-1}\,[F]$$

$$[k]^{-1} = \frac{1}{1.42EI}\begin{bmatrix} 1 & -0.5 \\ -0.5 & 1.67 \end{bmatrix}$$

$$\{\Delta\} = \frac{1}{1.42EI}\begin{bmatrix} 1 & -0.5 \\ -0.5 & 1.67 \end{bmatrix} \times \begin{Bmatrix} -50 \\ -40 \end{Bmatrix}$$

$$\{\Delta\} = \begin{Bmatrix} -21.13/EI \\ -29.44/EI \end{Bmatrix} \begin{matrix} \longrightarrow \theta_B \\ \longrightarrow \theta_C \end{matrix}$$

Step 9:

Get the element displacements $\{\delta\}$ using the equation,

$$\{\delta\} = [\beta]\{\Delta\}$$

$$\{\delta\} = \begin{bmatrix} 0 & 0 \\ 1 & 0 \\ 1 & 0 \\ 0 & 1 \end{bmatrix} \times \frac{1}{EI}\begin{Bmatrix} -21.13 \\ -29.44 \end{Bmatrix} = \frac{1}{EI}\begin{Bmatrix} 0 \\ -21.13 \\ -21.13 \\ -29.44 \end{Bmatrix}$$

Step 10:

Let the element forces in element coordinates be $\{p'\}$. Then

$$\{p'\} = [k][\delta]$$

$$\{p'\} = EI\begin{bmatrix} 0.67 & 0.33 & 0 & 0 \\ 0.33 & 0.67 & 0 & 0 \\ 0 & 0 & 1 & 0.5 \\ 0 & 0 & 0.5 & 1 \end{bmatrix} \times \frac{1}{EI}\begin{Bmatrix} 0 \\ -21.13 \\ -21.13 \\ -29.44 \end{Bmatrix} = \begin{Bmatrix} -6.97 \\ -14.16 \\ -35.85 \\ -40.01 \end{Bmatrix}$$

Step 11:

Compute the final forces

$$\{p^f\} = \{p^0\} + \{p'\}$$

$$\{p^f\} = \begin{Bmatrix} -90 \\ 90 \\ -40 \\ 40 \end{Bmatrix} + \begin{Bmatrix} -6.97 \\ -14.16 \\ -35.85 \\ -40.01 \end{Bmatrix} = \begin{Bmatrix} -96.97 \\ -75.85 \\ -75.85 \\ 0 \end{Bmatrix}$$

Final moments and bending moment diagrams as as shown in Fig. 3.2(f) and (g).

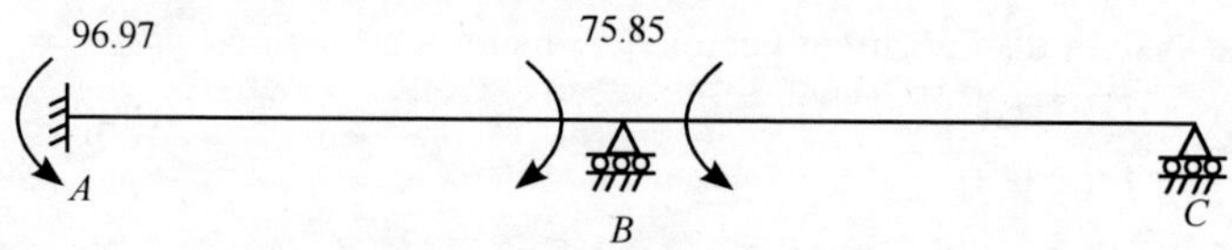

End moments

Fig. 3.2(f)

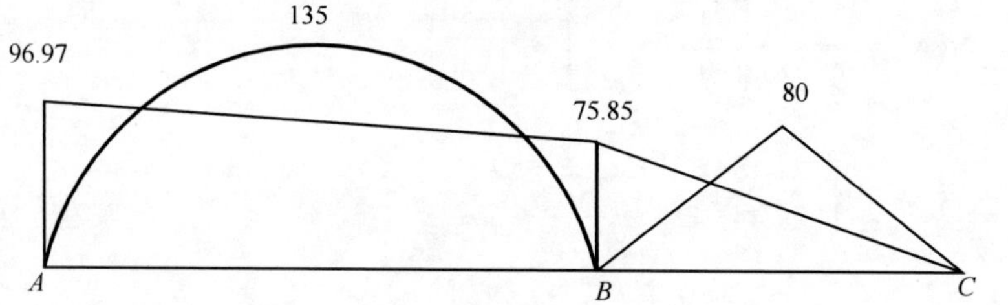

Final BMD

Fig. 3.2(g)

Example 3.3 Analyse the continuous beam loaded as shown in Fig. 3.3(a) by matrix stiffness method.

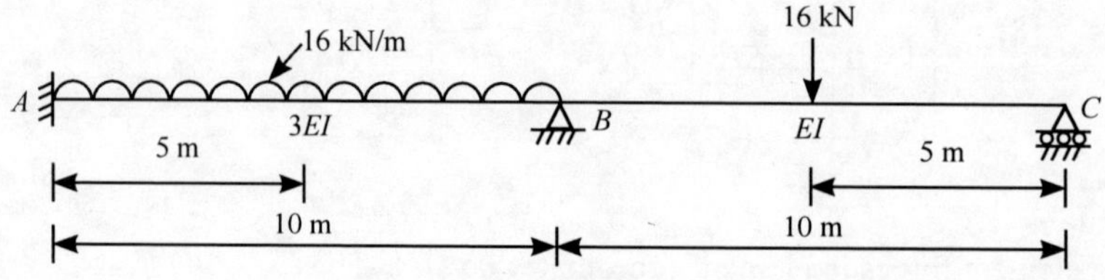

Fig. 3.3(a)

Solution:

Kinematic indeterminacy, $KI = 2$

θ_B and θ_C are system coordinates. Refer Fig. 3.3(b).

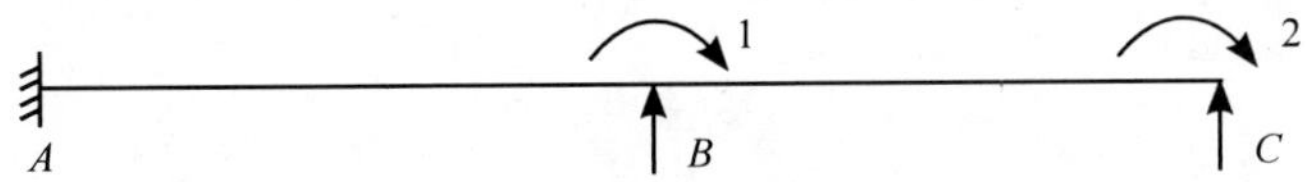

System coordinates

Fig. 3.3(b)

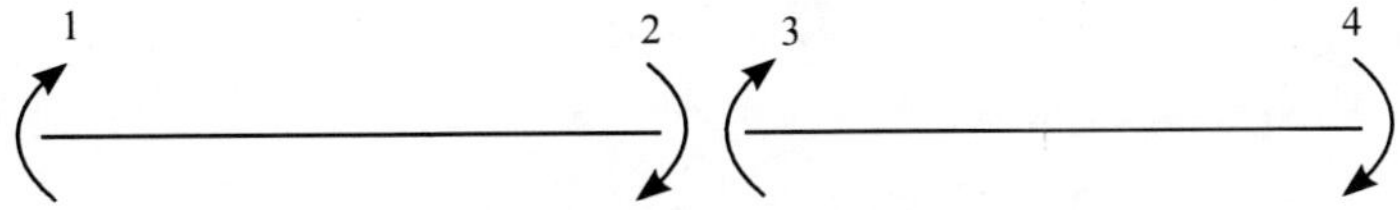

Element coordinates

Fig. 3.3(c)

Fixed End Moments

$$MF_{AB} = -\frac{wl^2}{12} = -133.33 \text{ kN-m}$$

$$MF_{BA} = \frac{wl^2}{12} = 133.33 \text{ kN-m}$$

$$MF_{BC} = -\frac{wl}{8} = -20 \text{ kN-m}$$

$$MF_{CB} = \frac{wl}{8} = 20 \text{ kN-m}$$

Forces in the Element Coordinates

$$\{P^0\} = \begin{Bmatrix} -133.33 \\ 133.33 \\ -20 \\ 20 \end{Bmatrix}$$

Transformation Matrix, [β]

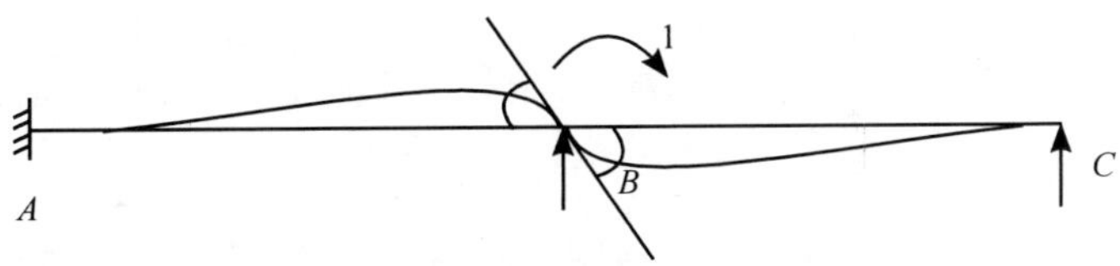

Fig. 3.3(d)

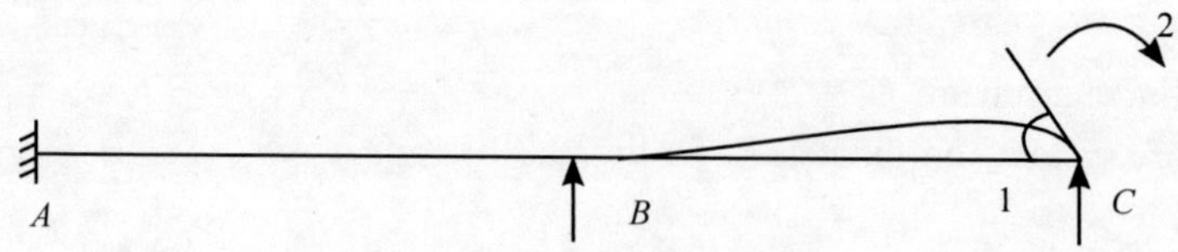

Fig. 3.3(e)

$$[\beta] = \begin{bmatrix} 0 & 0 \\ 1 & 0 \\ 1 & 0 \\ 0 & 1 \end{bmatrix}$$

Forces in the System Coordinates

$$\{F^0\} = [\beta]^T\{P^0\}$$

$$\{F^0\} = \begin{bmatrix} 0 & 1 & 1 & 0 \\ 0 & 0 & 0 & 1 \end{bmatrix} \times \begin{Bmatrix} -133.33 \\ 133.33 \\ -20 \\ 20 \end{Bmatrix} = \begin{Bmatrix} 113.33 \\ 20 \end{Bmatrix}$$

Let Applied Forces in System Coordinates be F'

$$\{F'\} = \begin{Bmatrix} 0 \\ 0 \end{Bmatrix}$$

System Forces

$$\{F\} = [\{F'\} - \{F^0\}]$$

$$\{F\} = \left[\begin{Bmatrix} 0 \\ 0 \end{Bmatrix} - \begin{Bmatrix} 113.33 \\ 20 \end{Bmatrix}\right] = \begin{Bmatrix} -113.33 \\ -20 \end{Bmatrix}$$

Element Stiffness Matrix

$$[k_1] = \frac{E_1 I_1}{l_1}\begin{bmatrix} 4 & 2 \\ 2 & 4 \end{bmatrix} = \frac{1}{10}\begin{bmatrix} 12 & 6 \\ 6 & 12 \end{bmatrix}$$

$$[k_2] = \frac{E_2 I_2}{l_2}\begin{bmatrix} 4 & 2 \\ 2 & 4 \end{bmatrix} = \frac{1}{10}\begin{bmatrix} 4 & 2 \\ 2 & 4 \end{bmatrix}$$

System Stiffness Matrix

$$[k] = [\beta]^T[k][\beta]$$

$$[k] = \begin{bmatrix} 0 & 1 & 1 & 0 \\ 0 & 0 & 0 & 1 \end{bmatrix} \times \frac{1}{10}\begin{bmatrix} 12 & 6 & 0 & 0 \\ 6 & 12 & 0 & 0 \\ 0 & 0 & 4 & 2 \\ 0 & 0 & 2 & 4 \end{bmatrix} \times \begin{bmatrix} 0 & 0 \\ 1 & 0 \\ 1 & 0 \\ 0 & 1 \end{bmatrix}$$

$$[k] = \frac{1}{10}\begin{bmatrix} 16 & 2 \\ 2 & 4 \end{bmatrix}$$

System Displacements

$$\{\Delta\} = [k]^{-1}[\{F'\} - \{F^0\}]$$

$$\{\Delta\} = [k]^{-1}\,[F]$$

$$[k]^{-1} = \frac{1}{6}\begin{bmatrix} 4 & -2 \\ -2 & 16 \end{bmatrix}$$

$$\{\Delta\} = \frac{1}{6}\begin{bmatrix} 4 & -2 \\ -2 & 16 \end{bmatrix} \times \begin{Bmatrix} -113.33 \\ -20 \end{Bmatrix}$$

$$\{\Delta\} = \begin{Bmatrix} -68.88 \\ -15.55 \end{Bmatrix} \begin{matrix} \longrightarrow \theta_B \\ \longrightarrow \theta_C \end{matrix}$$

Element Displacements

$$\{\Delta\} = [\beta]\{\Delta\}$$

$$\{\Delta\} = \begin{bmatrix} 0 & 0 \\ 1 & 0 \\ 1 & 0 \\ 0 & 1 \end{bmatrix} \times \begin{Bmatrix} -68.88 \\ -15.55 \end{Bmatrix} = \begin{Bmatrix} 0 \\ -68.88 \\ -68.88 \\ -15.55 \end{Bmatrix}$$

Forces along the Element Coordinates Consistent with the Displacements

$$\{p'\} = [k][\delta]$$

$$\{p'\} = \frac{1}{10}\begin{bmatrix} 12 & 6 & 0 & 0 \\ 6 & 12 & 0 & 0 \\ 0 & 0 & 4 & 2 \\ 0 & 0 & 2 & 4 \end{bmatrix} \times \begin{Bmatrix} 0 \\ -68.88 \\ -68.88 \\ -15.55 \end{Bmatrix} = \begin{Bmatrix} -41.328 \\ -82.656 \\ -30.662 \\ -20 \end{Bmatrix}$$

Final Forces

$$\{p^f\} = \{p^0\} + \{p'\}$$

$$\{p^f\} = \begin{Bmatrix} -133.33 \\ 133.33 \\ -20 \\ 20 \end{Bmatrix} + \begin{Bmatrix} -41.328 \\ -82.656 \\ -30.662 \\ -20 \end{Bmatrix} = \begin{Bmatrix} -174.65 \\ 50.67 \\ -50.67 \\ 0 \end{Bmatrix}$$

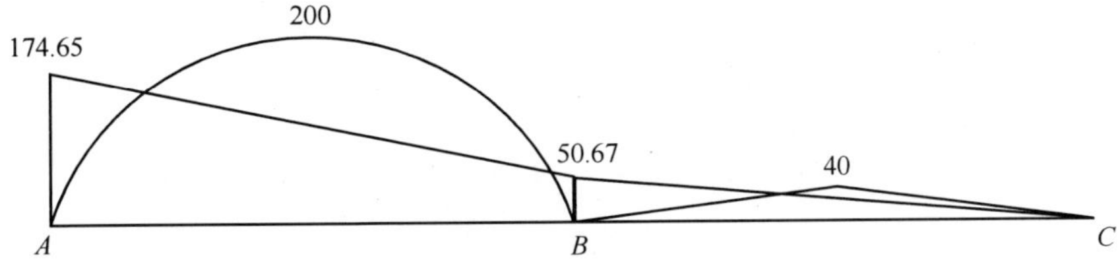

Final BMD

Fig. 3.3(f)

Example 3.4 Analyse the beam loaded as shown in Fig. 3.4(a) by matrix stiffness method.

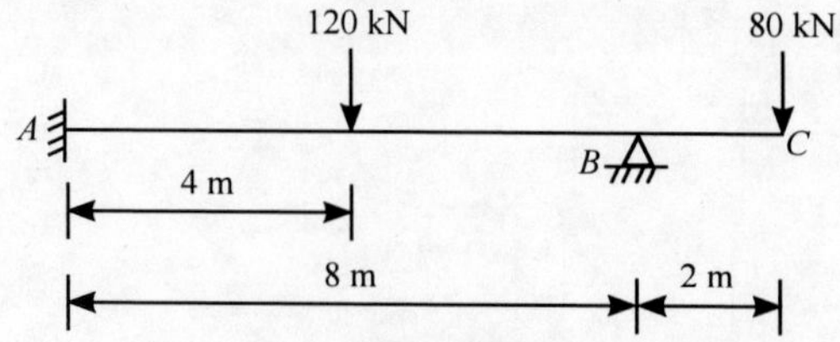

Fig. 3.4(a)

Solution:

Kinematic indeterminacy, $KI = 1$

θ_B is system coordinate

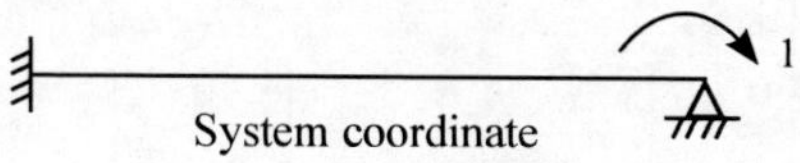

Fig. 3.4(b)

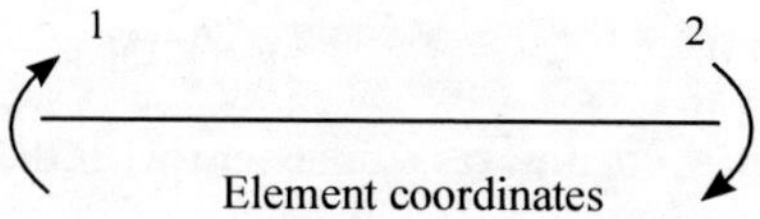

Fig. 3.4(c)

Fixed End Moments

$$MF_{AB} = -\frac{wl}{8} = -120 \text{ kN-m}$$

$$MF_{BA} = \frac{wl}{8} = 120 \text{ kN-m}$$

Forces in the Element Coordinates

$$\{P^0\} = \begin{Bmatrix} -120 \\ 120 \end{Bmatrix}$$

Transformation Matrix, [β]

$$[\beta] = \begin{bmatrix} 0 \\ 1 \end{bmatrix}$$

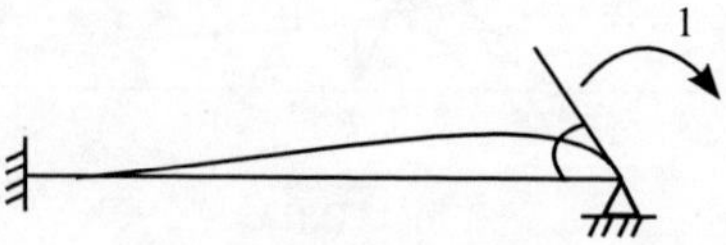

Fig. 3.4(d)

Forces in the System Coordinates

$$\{F^0\} = [\beta]^T\{P^0\}$$

$$\{F^0\} = \begin{bmatrix} 0 & 1 \end{bmatrix} \times \begin{Bmatrix} -120 \\ 120 \end{Bmatrix} = \{120\}$$

Forces Applied at System Coordinates

$$\{F'\} = \{160\}$$

System Forces

$$\{F\} = [\{F'\} - \{F^0\}]$$

$$\{F\} = \left[\{160\} - \{120\}\right] = \{40\}$$

Element Stiffness Matrix

$$[k] = \frac{EI}{l}\begin{bmatrix} 4 & 2 \\ 2 & 4 \end{bmatrix} = \frac{EI}{8}\begin{bmatrix} 4 & 2 \\ 2 & 4 \end{bmatrix}$$

System Stiffness Matrix

$$[k] = [\beta]^T[k][\beta]$$

$$[k] = \begin{bmatrix} 0 & 1 \end{bmatrix} \times \frac{EI}{8}\begin{bmatrix} 4 & 2 \\ 2 & 4 \end{bmatrix} \times \begin{bmatrix} 0 \\ 1 \end{bmatrix}$$

$$[k] = \frac{EI}{2}$$

$$[k]^{-1} = \frac{2}{EI}$$

System Displacements

$$\{\Delta\} = [k]^{-1}\ [\{F'\} - \{F^0\}]$$

$$\{\Delta\} = [k]^{-1}\ [F]$$

$$[k]^{-1} = \frac{2}{EI}$$

$$\{\Delta\} = \frac{2}{EI} \times \{40\} = \frac{80}{EI} \longrightarrow \theta_B$$

Element Displacements

$$\{\delta\} = [\beta]\{\Delta\}$$

$$\{\delta\} = \begin{bmatrix} 0 \\ 1 \end{bmatrix} \times \begin{Bmatrix} \dfrac{80}{EI} \end{Bmatrix} = \begin{Bmatrix} 0 \\ \dfrac{80}{EI} \end{Bmatrix}$$

Forces along the element coordinates consistent with the displacements,

$$\{p'\} = [k][\delta]$$

$$\{p'\} = \frac{EI}{8}\begin{bmatrix} 4 & 2 \\ 2 & 4 \end{bmatrix} \times \begin{Bmatrix} 0 \\ \frac{80}{EI} \end{Bmatrix} = \begin{Bmatrix} 20 \\ 40 \end{Bmatrix}$$

Final Forces

$$\{p^f\} = \{p^0\} + \{p'\}$$

$$\{p^f\} = \begin{Bmatrix} -120 \\ 120 \end{Bmatrix} + \begin{Bmatrix} 20 \\ 40 \end{Bmatrix} = \begin{Bmatrix} -100 \\ 160 \end{Bmatrix}$$

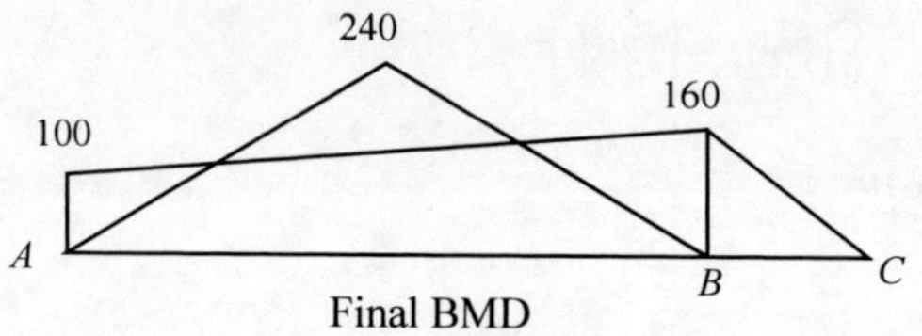

Fig. 3.4(e)

Example 3.5 Analyse the continuous beam loaded as shown in Fig. 3.5(a) by matrix stiffness method.

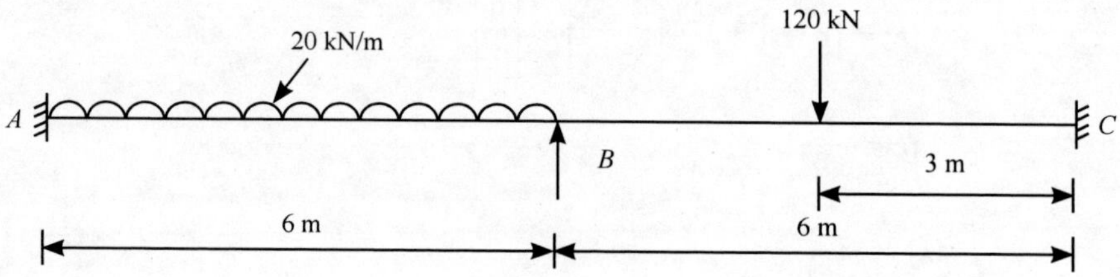

Fig. 3.5(a)

Solution:

Kinematic indeterminacy, $KI = 1$

θ_B is system coordinate

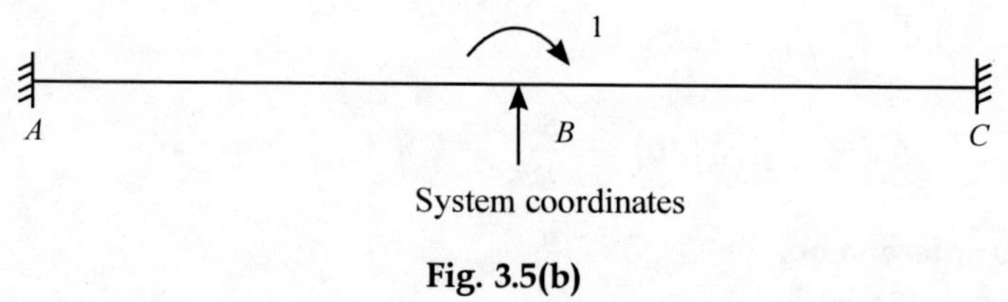

Fig. 3.5(b)

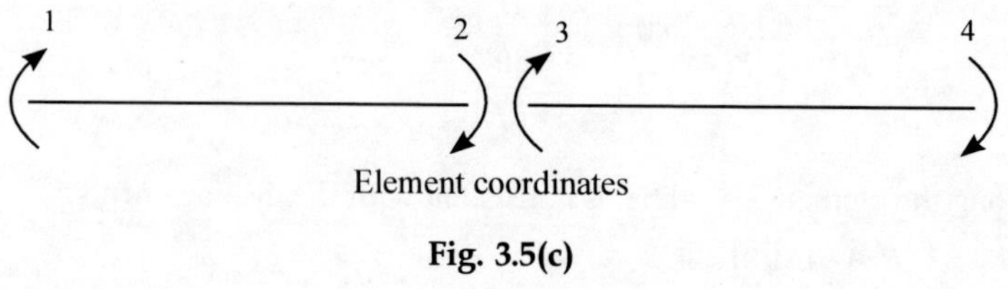

Fig. 3.5(c)

Fixed End Moments

$$MF_{AB} = -\frac{wl^2}{12} = -60 \text{ kN-m}$$

$$MF_{BA} = \frac{wl^2}{12} = 60 \text{ kN-m}$$

$$MF_{BC} = -\frac{wl}{8} = -90 \text{ kN-m}$$

$$MF_{CB} = \frac{wl}{8} = 90 \text{ kN-m}$$

Forces in the Element Coordinates

$$\{P^0\} = \begin{Bmatrix} -60 \\ 60 \\ -90 \\ 90 \end{Bmatrix}$$

Transformation Matrix, [β]

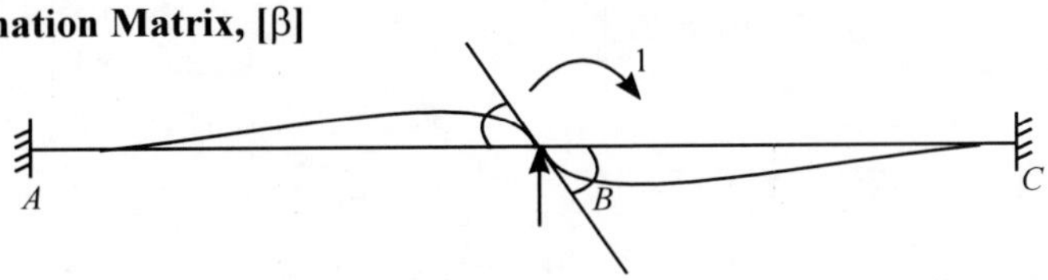

Fig. 3.5(d)

$$[\beta] = \begin{bmatrix} 0 \\ 1 \\ 1 \\ 0 \end{bmatrix}$$

Forces in the System Coordinates

$$\{F^0\} = [\beta]^T \{P^0\}$$

$$\{F^0\} = \begin{bmatrix} 0 & 1 & 1 & 0 \end{bmatrix} \times \begin{Bmatrix} -60 \\ 60 \\ -90 \\ 90 \end{Bmatrix} = \{-30\}$$

Forces applied at system coordinates

$$\{F'\} = \{0\}$$

System Forces

$$\{F\} = [\{F'\} - \{F^0\}]$$

$$\{F\} = \left[\{0\} - \{-30\}\right] = \{30\}$$

Element Stiffness Matrix

$$[k_1] = \frac{E_1 I_1}{l_1}\begin{bmatrix} 4 & 2 \\ 2 & 4 \end{bmatrix} = \frac{EI}{6}\begin{bmatrix} 4 & 2 \\ 2 & 4 \end{bmatrix} = EI\begin{bmatrix} 0.67 & 0.33 \\ 0.33 & 0.67 \end{bmatrix}$$

$$[k_1] = [k_2] = EI\begin{bmatrix} 0.67 & 0.33 \\ 0.33 & 0.67 \end{bmatrix}$$

Assembled Element Stiffness Matrix, [*k*]

$$[k] = EI\begin{bmatrix} 0.67 & 0.33 & 0 & 0 \\ 0.33 & 0.67 & 0 & 0 \\ 0 & 0 & 0.67 & 0.33 \\ 0 & 0 & 0.33 & 0.67 \end{bmatrix}$$

System Stiffness Matrix

$$[k] = [\beta]^T[k][\beta]$$

$$[k] = \begin{bmatrix} 0 & 1 & 1 & 0 \end{bmatrix} \times EI\begin{bmatrix} 0.67 & 0.33 & 0 & 0 \\ 0.33 & 0.67 & 0 & 0 \\ 0 & 0 & 0.67 & 0.33 \\ 0 & 0 & 0.33 & 0.67 \end{bmatrix} \times \begin{bmatrix} 0 \\ 1 \\ 1 \\ 0 \end{bmatrix}$$

$$[k] = 1.34\,EI$$

$$[k]^{-1} = \frac{1}{1.34EI}$$

System Displacements

$$\{\Delta\} = [k]^{-1}[\{F'\} - \{F^0\}]$$

$$\{\Delta\} = [k]^{-1}\,[F]$$

$$[k]^{-1} = \frac{1}{1.34EI}$$

$$\{\Delta\} = \frac{1}{1.34EI} \times \{30\} = \frac{22.39}{EI}$$

Element Displacements

$$\{\delta\} = [\beta]\{\Delta\}$$

$$\{\delta\} = \begin{bmatrix} 0 \\ 1 \\ 1 \\ 0 \end{bmatrix} \times \left\{\frac{22.39}{EI}\right\} = \frac{1}{EI}\begin{Bmatrix} 0 \\ 22.39 \\ 22.39 \\ 0 \end{Bmatrix}$$

Forces along the element coordinates consistent with the displacements,

$$\{p'\} = [k][\delta]$$

$$\{p'\} = EI\begin{bmatrix} 0.67 & 0.33 & 0 & 0 \\ 0.33 & 0.67 & 0 & 0 \\ 0 & 0 & 0.67 & 0.33 \\ 0 & 0 & 0.33 & 0.67 \end{bmatrix} \times \frac{1}{EI}\begin{Bmatrix} 0 \\ 22.39 \\ 22.39 \\ 0 \end{Bmatrix} = \begin{Bmatrix} 7.39 \\ 15 \\ 15 \\ 7.39 \end{Bmatrix}$$

Final Forces

$$\{p^f\} = \{p^0\} + \{p'\}$$

$$\{p^f\} = \begin{Bmatrix} -60 \\ 60 \\ -90 \\ 90 \end{Bmatrix} + \begin{Bmatrix} 7.39 \\ 15 \\ 15 \\ 7.39 \end{Bmatrix} = \begin{Bmatrix} -52.61 \\ 75 \\ -75 \\ 97.39 \end{Bmatrix}$$

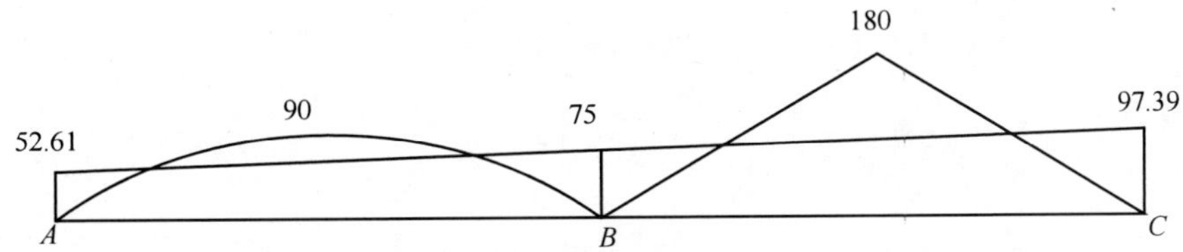

Final BMD

Fig. 3.5(e)

3.3 ANALYSIS OF PLANE RIGID FRAMES

Plane rigid frames are analysed here by stiffness matrix method.

Example 3.6 Analyze the portal frame as shown in Fig. 3.6(a) by stiffness matrix method.

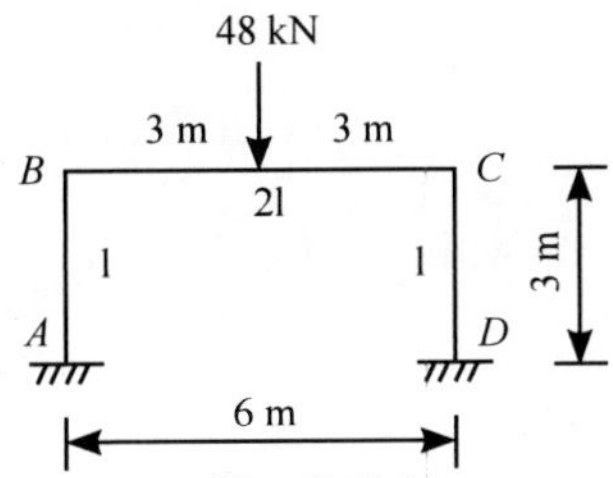

Fig. 3.6(a)

Solution:

Kinematic indeterminacy, $KI = 2$

θ_B and θ_C are possible displacements

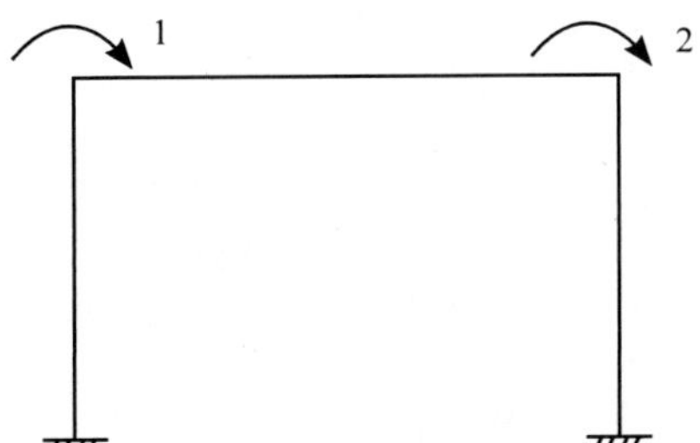

System coordinates

Fig. 3.6(b)

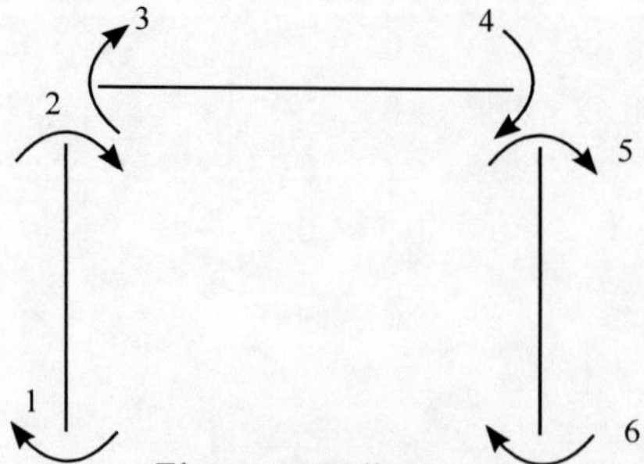

Element coordinates

Fig. 3.6(c)

Fixed End Moments

$$MF_{AB} = MF_{BA} = MF_{CD} = MF_{DC} = 0$$

$$MF_{BC} = -\frac{wl}{8} = -36 \text{ kN-m}$$

$$MF_{CB} = \frac{wl}{8} = 36 \text{ kN-m}$$

Forces in the Element Coordinates

$$\{P^0\} = \begin{Bmatrix} 0 \\ 0 \\ -36 \\ 36 \\ 0 \\ 0 \end{Bmatrix}$$

Transformation Matrix, [β]

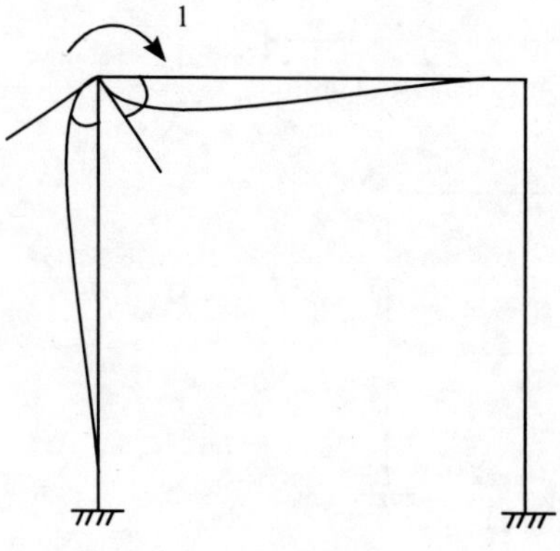

Fig. 3.6(d)

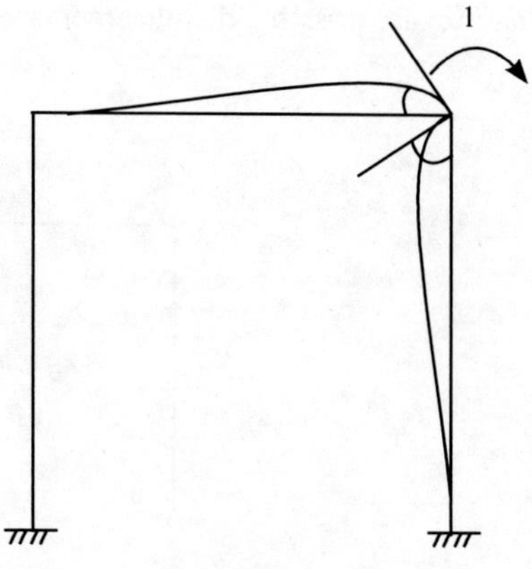

Fig. 3.6(e)

$$[\beta] = \begin{bmatrix} 0 & 0 \\ 1 & 0 \\ 1 & 0 \\ 0 & 1 \\ 0 & 1 \\ 0 & 0 \end{bmatrix}$$

Forces in the system coordinates corresponding to $\{P^0\}$, i.e., fixed coordinate forces,

$$\{F^0\} = [\beta]^T \{P^0\}$$

$$\{F^0\} = \begin{bmatrix} 0 & 1 & 1 & 0 & 0 & 0 \\ 0 & 0 & 0 & 1 & 1 & 0 \end{bmatrix} \times \begin{Bmatrix} 0 \\ 0 \\ -36 \\ 36 \\ 0 \\ 0 \end{Bmatrix} = \begin{Bmatrix} -36 \\ 36 \end{Bmatrix}$$

Forces Applied at System Coordinates

$$\{F'\} = \begin{Bmatrix} 0 \\ 0 \end{Bmatrix}$$

System Forces

$$\{F\} = [\{F'\} - \{F^0\}]$$

$$\{F\} = \left[\begin{Bmatrix} 0 \\ 0 \end{Bmatrix} - \begin{Bmatrix} -36 \\ 36 \end{Bmatrix} \right] = \begin{Bmatrix} 36 \\ -36 \end{Bmatrix}$$

Element Stiffness Matrix

$$[k_1] = \frac{E_1 I_1}{l_1} \begin{bmatrix} 4 & 2 \\ 2 & 4 \end{bmatrix} = \frac{EI}{3} \begin{bmatrix} 4 & 2 \\ 2 & 4 \end{bmatrix} = [k_3]$$

$$[k_2] = \frac{E_2 I_2}{l_2} \begin{bmatrix} 4 & 2 \\ 2 & 4 \end{bmatrix} = \frac{E(2I)}{6} \begin{bmatrix} 4 & 2 \\ 2 & 4 \end{bmatrix} = \frac{EI}{3} \begin{bmatrix} 4 & 2 \\ 2 & 4 \end{bmatrix}$$

Assembled element stiffness matrix, $[k]$

$$[k] = \frac{EI}{3} \begin{bmatrix} 4 & 2 & 0 & 0 & 0 & 0 \\ 2 & 4 & 0 & 0 & 0 & 0 \\ 0 & 0 & 4 & 2 & 0 & 0 \\ 0 & 0 & 2 & 4 & 0 & 0 \\ 0 & 0 & 0 & 0 & 4 & 2 \\ 0 & 0 & 0 & 0 & 2 & 4 \end{bmatrix}$$

System Stiffness Matrix

$$[k] = [\beta]^T[k][\beta]$$

$$[k] = \begin{bmatrix} 0 & 1 & 1 & 0 & 0 & 0 \\ 0 & 0 & 0 & 1 & 1 & 0 \end{bmatrix} \times \frac{EI}{3}\begin{bmatrix} 4 & 2 & 0 & 0 & 0 & 0 \\ 2 & 4 & 0 & 0 & 0 & 0 \\ 0 & 0 & 4 & 2 & 0 & 0 \\ 0 & 0 & 2 & 4 & 0 & 0 \\ 0 & 0 & 0 & 0 & 4 & 2 \\ 0 & 0 & 0 & 0 & 2 & 4 \end{bmatrix} \times \begin{bmatrix} 0 & 0 \\ 1 & 0 \\ 1 & 0 \\ 0 & 1 \\ 0 & 1 \\ 0 & 0 \end{bmatrix}$$

$$[k] = \frac{EI}{3}\begin{bmatrix} 8 & 2 \\ 2 & 8 \end{bmatrix}$$

System Displacements

$$\{\Delta\} = [k]^{-1}[\{F'\} - \{F^0\}]$$

$$\{\Delta\} = [k]^{-1}\,[F]$$

$$[k]^{-1} = \frac{1}{20EI}\begin{bmatrix} 8 & -2 \\ -2 & 8 \end{bmatrix}$$

$$\{\Delta\} = \frac{1}{20EI}\begin{bmatrix} 8 & -2 \\ -2 & 8 \end{bmatrix} \times \begin{Bmatrix} 36 \\ -36 \end{Bmatrix}$$

$$\{\Delta\} = \frac{1}{EI}\begin{Bmatrix} 18 \\ -18 \end{Bmatrix} \begin{matrix} \theta_B \\ \theta_C \end{matrix}$$

Element Displacements

$$\{\delta\} = [\beta]\{\Delta\}$$

$$\{\delta\} = \begin{bmatrix} 0 & 0 \\ 1 & 0 \\ 1 & 0 \\ 0 & 1 \\ 0 & 1 \\ 0 & 0 \end{bmatrix} \times \begin{Bmatrix} \dfrac{18}{EI} \\ \dfrac{-18}{EI} \end{Bmatrix} = \begin{Bmatrix} 0 \\ \dfrac{18}{EI} \\ \dfrac{18}{EI} \\ \dfrac{-18}{EI} \\ \dfrac{-18}{EI} \\ 0 \end{Bmatrix}$$

Forces along the element coordinates consistent with the displacements

$$\{p'\} = [k][\Delta]$$

$$\{p'\} = \frac{EI}{3}\begin{bmatrix} 4 & 2 & 0 & 0 & 0 & 0 \\ 2 & 4 & 0 & 0 & 0 & 0 \\ 0 & 0 & 4 & 2 & 0 & 0 \\ 0 & 0 & 2 & 4 & 0 & 0 \\ 0 & 0 & 0 & 0 & 4 & 2 \\ 0 & 0 & 0 & 0 & 2 & 4 \end{bmatrix} \times \frac{1}{EI}\begin{Bmatrix} 0 \\ 18 \\ 18 \\ -18 \\ -18 \\ 0 \end{Bmatrix} = \begin{Bmatrix} 12 \\ 24 \\ 12 \\ -12 \\ -24 \\ -12 \end{Bmatrix}$$

Final Forces

$$\{p^f\} = \{p^0\} + \{p'\}$$

$$\{p^f\} = \begin{Bmatrix} 0 \\ 0 \\ -36 \\ 36 \\ 0 \\ 0 \end{Bmatrix} + \begin{Bmatrix} 12 \\ 24 \\ 12 \\ -12 \\ -24 \\ -12 \end{Bmatrix} = \begin{Bmatrix} 12 \\ 24 \\ -24 \\ 24 \\ -24 \\ -12 \end{Bmatrix}$$

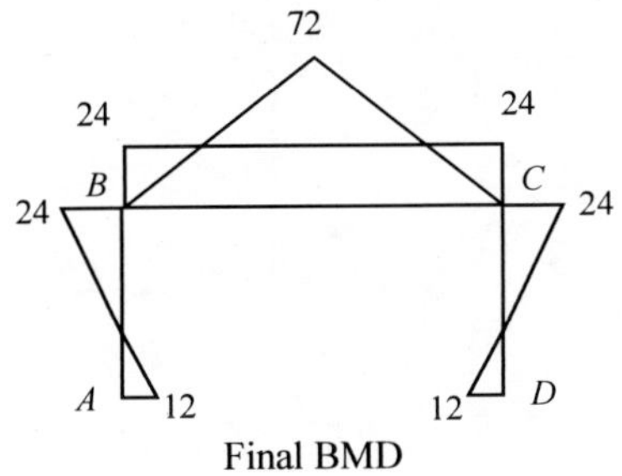

Fig. 3.6(f)

Example 3.7 Analyze the portal frame as shown in Fig. 3.7(a) by stiffness matrix method.

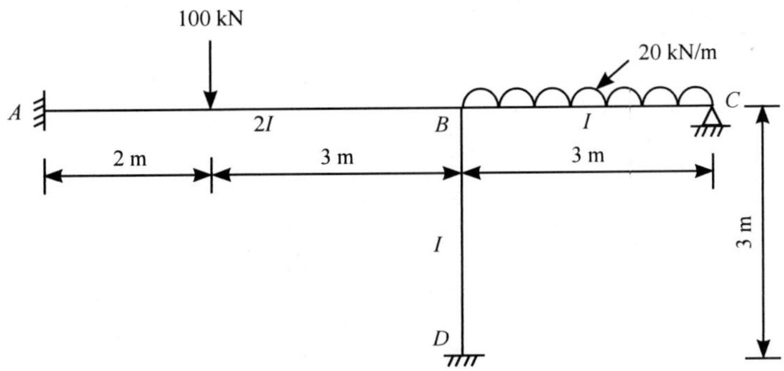

Fig. 3.7(a)

Solution:

Kinematic indeterminacy, $KI = 2$

θ_B and θ_C are possible displacements

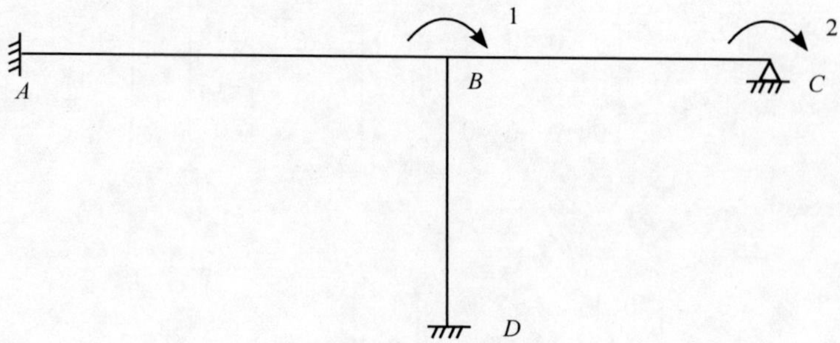

System coordinates

Fig. 3.7(b)

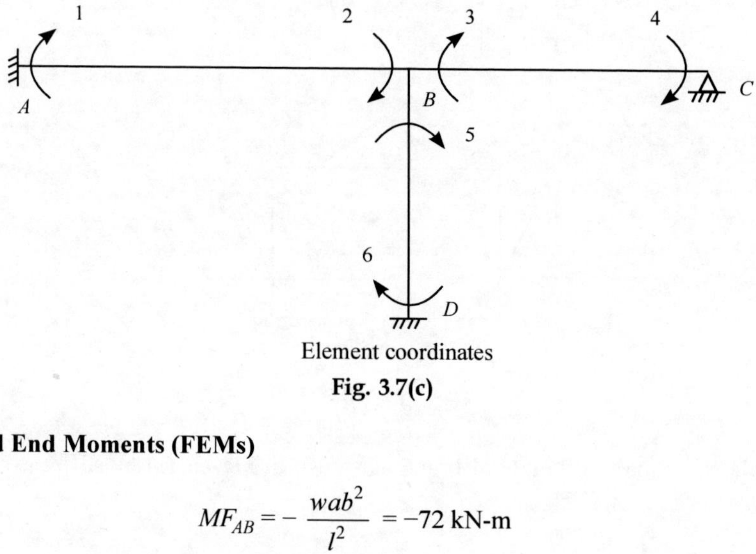

Element coordinates

Fig. 3.7(c)

Fixed End Moments (FEMs)

$$MF_{AB} = -\frac{wab^2}{l^2} = -72 \text{ kN-m}$$

$$MF_{BA} = \frac{wa^2b}{l^2} = 48 \text{ kN-m}$$

$$MF_{BC} = -\frac{wl^2}{12} = -15 \text{ kN-m}$$

$$MF_{CB} = \frac{wl^2}{12} = 15 \text{ kN-m}$$

$$MF_{BD} = MF_{DB} = 0$$

Forces in the Element Coordinates

$$\{P^0\} = \begin{Bmatrix} -72 \\ 48 \\ -15 \\ 15 \\ 0 \\ 0 \end{Bmatrix}$$

Transformation Matrix, [β]

$$[\beta] = \begin{bmatrix} 0 & 0 \\ 1 & 0 \\ 1 & 0 \\ 0 & 1 \\ 1 & 0 \\ 0 & 0 \end{bmatrix}$$

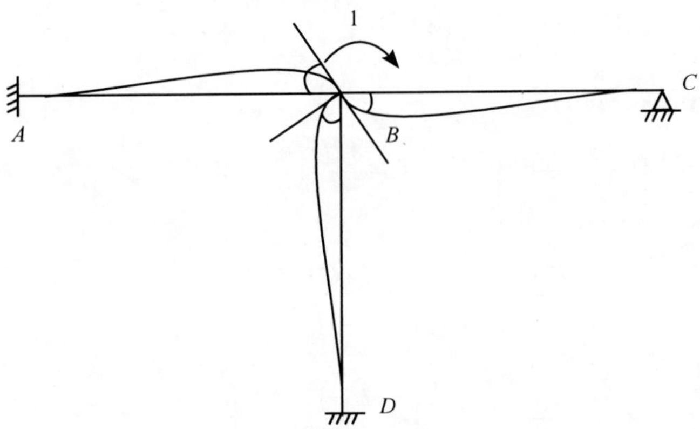

Fig. 3.7(d)

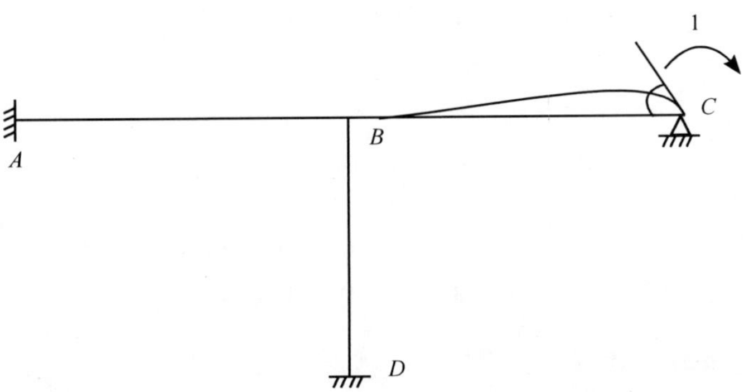

Fig. 3.7(e)

Forces in the system coordinates corresponding to $\{P^0\}$, i.e., fixed coordinate forces,

$\{F^0\} = [\beta]^T\{P^0\}$

$$\{F^0\} = \begin{bmatrix} 0 & 1 & 1 & 0 & 1 & 0 \\ 0 & 0 & 0 & 1 & 0 & 0 \end{bmatrix} \times \begin{Bmatrix} -72 \\ 48 \\ -15 \\ 15 \\ 0 \\ 0 \end{Bmatrix} = \begin{Bmatrix} 33 \\ 15 \end{Bmatrix}$$

Forces Applied at System Coordinates

$$\{F'\} = \begin{Bmatrix} 0 \\ 0 \end{Bmatrix}$$

System Forces

$$\{F\} = [\{F'\} - \{F^0\}]$$

$$\{F\} = \left[\begin{Bmatrix} 0 \\ 0 \end{Bmatrix} - \begin{Bmatrix} 33 \\ 15 \end{Bmatrix}\right] = \begin{Bmatrix} -33 \\ -15 \end{Bmatrix}$$

Element Stiffness Matrix

$$[k_1] = \frac{E_1 I_1}{l_1}\begin{bmatrix} 4 & 2 \\ 2 & 4 \end{bmatrix} = \frac{E(2I)}{5}\begin{bmatrix} 4 & 2 \\ 2 & 4 \end{bmatrix}$$

$$[k_2] = \frac{E_2 I_2}{l_2}\begin{bmatrix} 4 & 2 \\ 2 & 4 \end{bmatrix} = \frac{E(I)}{3}\begin{bmatrix} 4 & 2 \\ 2 & 4 \end{bmatrix}$$

$$[k_3] = \frac{E_3 I_3}{l_3}\begin{bmatrix} 4 & 2 \\ 2 & 4 \end{bmatrix} = \frac{E(I)}{3}\begin{bmatrix} 4 & 2 \\ 2 & 4 \end{bmatrix}$$

Assembled Element Stiffness Matrix, [*k*]

$$[k] = \frac{EI}{15}\begin{bmatrix} 24 & 12 & 0 & 0 & 0 & 0 \\ 12 & 24 & 0 & 0 & 0 & 0 \\ 0 & 0 & 20 & 10 & 0 & 0 \\ 0 & 0 & 10 & 20 & 0 & 0 \\ 0 & 0 & 0 & 0 & 20 & 10 \\ 0 & 0 & 0 & 0 & 10 & 20 \end{bmatrix}$$

System Stiffness Matrix

$$[k] = [\beta]^T[k][\beta]$$

$$[k] = \begin{bmatrix} 0 & 1 & 1 & 0 & 1 & 0 \\ 0 & 0 & 0 & 1 & 0 & 0 \end{bmatrix} \times \frac{EI}{15} \begin{bmatrix} 24 & 12 & 0 & 0 & 0 & 0 \\ 12 & 24 & 0 & 0 & 0 & 0 \\ 0 & 0 & 20 & 10 & 0 & 0 \\ 0 & 0 & 10 & 20 & 0 & 0 \\ 0 & 0 & 0 & 0 & 20 & 10 \\ 0 & 0 & 0 & 0 & 10 & 20 \end{bmatrix} \times \begin{bmatrix} 0 & 0 \\ 1 & 0 \\ 1 & 0 \\ 0 & 1 \\ 1 & 0 \\ 0 & 0 \end{bmatrix}$$

$$[k] = \frac{EI}{15} \begin{bmatrix} 64 & 10 \\ 10 & 20 \end{bmatrix}$$

System Displacements

$$\{\Delta\} = [k]^{-1}[\{F'\} - \{F^0\}]$$

$$\{\Delta\} = [k]^{-1}\,[F]$$

$$[k]^{-1} = \frac{15}{EI} \begin{bmatrix} \frac{1}{59} & -\frac{1}{118} \\ -\frac{1}{118} & \frac{16}{295} \end{bmatrix}$$

$$\{\Delta\} = \frac{15}{EI} \begin{bmatrix} \frac{1}{59} & -\frac{1}{118} \\ -\frac{1}{118} & \frac{16}{295} \end{bmatrix} \times \begin{Bmatrix} -33 \\ -15 \end{Bmatrix}$$

$$\{\Delta\} = \frac{15}{EI} \begin{Bmatrix} -\frac{51}{118} \\ -\frac{63}{118} \end{Bmatrix} \begin{matrix} \longrightarrow \theta_B \\ \longrightarrow \theta_C \end{matrix}$$

Element Forces

$$\{p'\} = [k][\delta]$$

$$\{\delta\} = [\beta]\{\Delta\}$$

$$\{p'\} = [k]\,[\beta]\{\Delta\}$$

$$\{p'\} = \frac{EI}{15} \begin{bmatrix} 24 & 12 & 0 & 0 & 0 & 0 \\ 12 & 24 & 0 & 0 & 0 & 0 \\ 0 & 0 & 20 & 10 & 0 & 0 \\ 0 & 0 & 10 & 20 & 0 & 0 \\ 0 & 0 & 0 & 0 & 20 & 10 \\ 0 & 0 & 0 & 0 & 10 & 20 \end{bmatrix} \times \begin{bmatrix} 0 & 0 \\ 1 & 0 \\ 1 & 0 \\ 0 & 1 \\ 1 & 0 \\ 0 & 0 \end{bmatrix} \times \frac{15}{EI} \begin{Bmatrix} -\frac{51}{118} \\ -\frac{63}{118} \end{Bmatrix} = \begin{Bmatrix} -5.186 \\ -10.37 \\ -13.986 \\ -15 \\ -8.644 \\ -4.32 \end{Bmatrix}$$

Final Forces

$$\{p^f\} = \{p^0\} + \{p'\}$$

$$\{p^f\} = \begin{Bmatrix} -72 \\ 48 \\ -15 \\ 15 \\ 0 \\ 0 \end{Bmatrix} + \begin{Bmatrix} -5.186 \\ -10.37 \\ -13.986 \\ -15 \\ -8.644 \\ -4.32 \end{Bmatrix} = \begin{Bmatrix} -77.186 \\ 37.627 \\ -28.983 \\ 0 \\ -8.644 \\ -4.322 \end{Bmatrix}$$

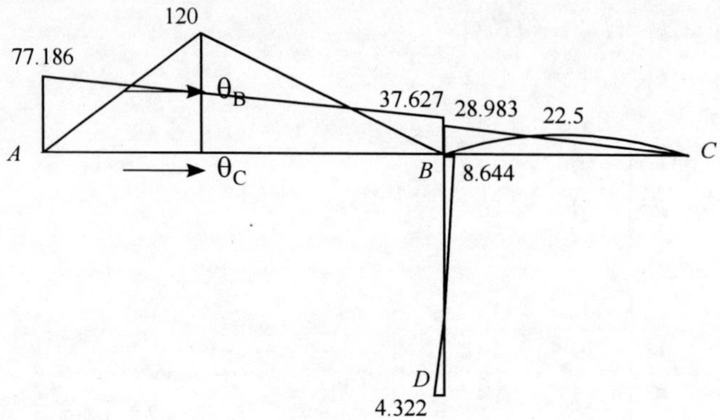

Final BMD

Fig. 3.7(f)

Example 3.8 Analyze the frame as shown in Fig. 3.8(a) by stiffness matrix method.

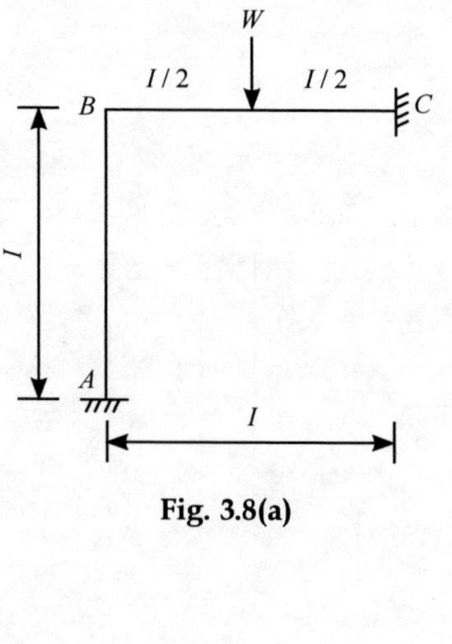

Fig. 3.8(a)

Solution:

Kinematic indeterminacy, $KI = 1$

θ_B is system coordinate

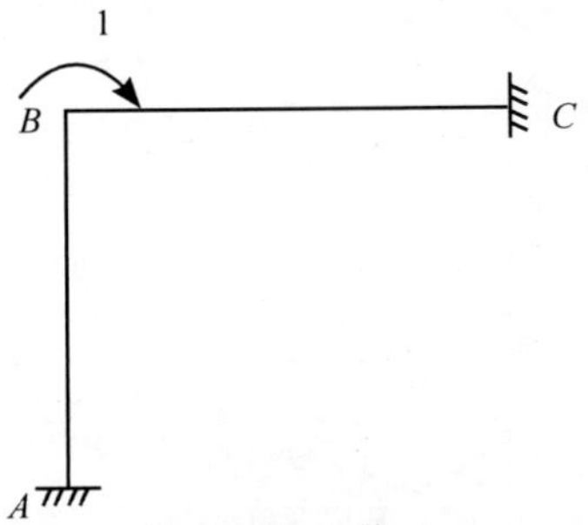

System coordinates

Fig. 3.8(b)

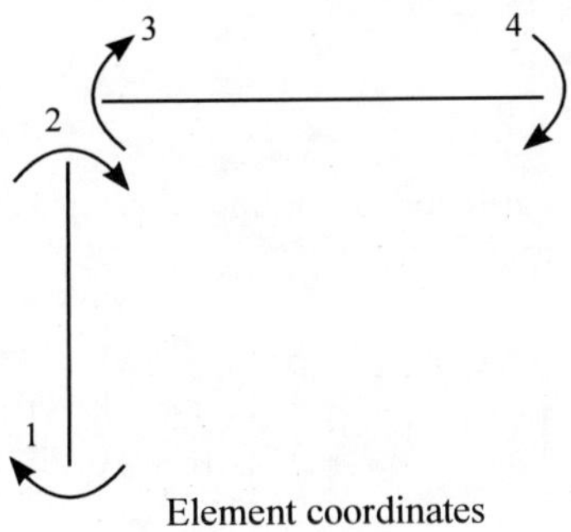

Element coordinates

Fig. 3.8(c)

Fixed End Moments (FEMs)

$$MF_{AB} = MF_{BA} = 0$$

$$MF_{BC} = -\frac{Wl}{8}$$

$$MF_{CB} = \frac{Wl}{8}$$

$$MF_{BD} = MF_{DB} = 0$$

Forces in the Element Coordinates

$$\{P^0\} = \begin{Bmatrix} 0 \\ 0 \\ -Wl/8 \\ Wl/8 \end{Bmatrix}$$

Transformation Matrix, [β]

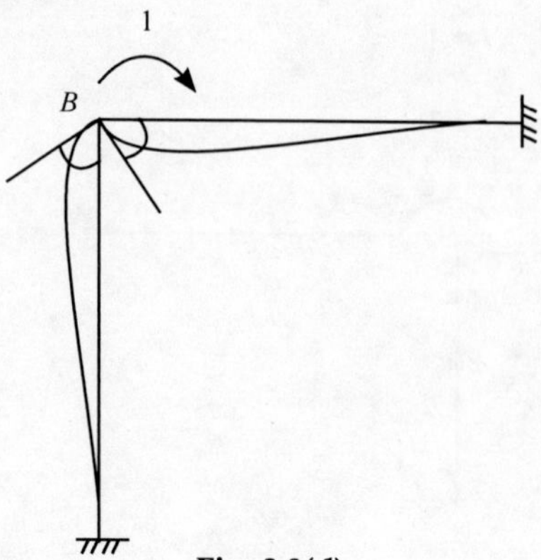

Fig. 3.8(d)

$$[\beta] = \begin{bmatrix} 0 \\ 1 \\ 1 \\ 0 \end{bmatrix}$$

Forces in the system coordinates corresponding to $\{P^0\}$, i.e., fixed coordinate forces,

$$\{F^0\} = [\beta]^T\{P^0\}$$

$$\{F^0\} = \begin{bmatrix} 0 & 1 & 1 & 0 \end{bmatrix} \times \begin{Bmatrix} 0 \\ 0 \\ -Wl/8 \\ Wl/8 \end{Bmatrix} = \{-Wl/8\}$$

Forces Applied at System Coordinates

$$\{F'\} = \{0\}$$

System Forces

$$\{F\} = [\{F'\} - \{F^0\}]$$

$$\{F\} = \left[\{0\} - \{-Wl/8\}\right] = \{Wl/8\}$$

Element Stiffness Matrix

$$[k_1] = \frac{EI}{l}\begin{bmatrix} 4 & 2 \\ 2 & 4 \end{bmatrix} = [k_2]$$

Assembled Element Stiffness Matrix, [*k*]

$$[k] = \frac{EI}{l}\begin{bmatrix} 4 & 2 & 0 & 0 \\ 2 & 4 & 0 & 0 \\ 0 & 0 & 4 & 2 \\ 0 & 0 & 2 & 4 \end{bmatrix}$$

System Stiffness Matrix

$$[k] = [\beta]^T[k][\beta]$$

$$[k] = \begin{bmatrix} 0 & 1 & 1 & 0 \end{bmatrix} \times \frac{EI}{l}\begin{bmatrix} 4 & 2 & 0 & 0 \\ 2 & 4 & 0 & 0 \\ 0 & 0 & 4 & 2 \\ 0 & 0 & 2 & 4 \end{bmatrix} \times \begin{bmatrix} 0 \\ 1 \\ 1 \\ 0 \end{bmatrix}$$

$$[k] = \frac{EI}{l}\,[8]$$

$$[k]^{-1} = \frac{l}{8EI}$$

System Displacements

$$\{\Delta\} = [k]^{-1}[\{F'\} - \{F^0\}]$$

$$\{\Delta\} = [k]^{-1}\,[F]$$

$$[k]^{-1} = \frac{l}{8EI}$$

$$\{\Delta\} = \frac{l}{8EI} \times \{Wl/8\} = \frac{Wl^2}{64EI} \rightarrow \theta_B$$

Element Displacements

$$\{\delta\} = [\beta]\{\Delta\}$$

$$\{\delta\} = \begin{bmatrix} 0 \\ 1 \\ 1 \\ 0 \end{bmatrix} \times \left\{\frac{Wl^2}{64EI}\right\} = \left\{\frac{Wl^2}{64EI}\right\}\begin{Bmatrix} 0 \\ 1 \\ 1 \\ 0 \end{Bmatrix}$$

Forces along the element coordinates consistent with the displacements

$$\{p'\} = [k][\delta]$$

$$\{p'\} = \frac{EI}{l}\begin{bmatrix} 4 & 2 & 0 & 0 \\ 2 & 4 & 0 & 0 \\ 0 & 0 & 4 & 2 \\ 0 & 0 & 2 & 4 \end{bmatrix} \times \left\{\frac{Wl^2}{64EI}\right\}\begin{Bmatrix} 0 \\ 1 \\ 1 \\ 0 \end{Bmatrix} = \frac{Wl}{64}\begin{Bmatrix} 2 \\ 4 \\ 4 \\ 2 \end{Bmatrix}$$

Final Forces

$$\{p^f\} = \{p^0\} + \{p'\}$$

$$\{p^f\} = \begin{Bmatrix} 0 \\ 0 \\ -Wl/8 \\ Wl/8 \end{Bmatrix} + \frac{Wl}{64}\begin{Bmatrix} 2 \\ 4 \\ 4 \\ 2 \end{Bmatrix} = Wl\begin{bmatrix} 0.03125 \\ 0.0625 \\ -0.0625 \\ 0.15625 \end{bmatrix}$$

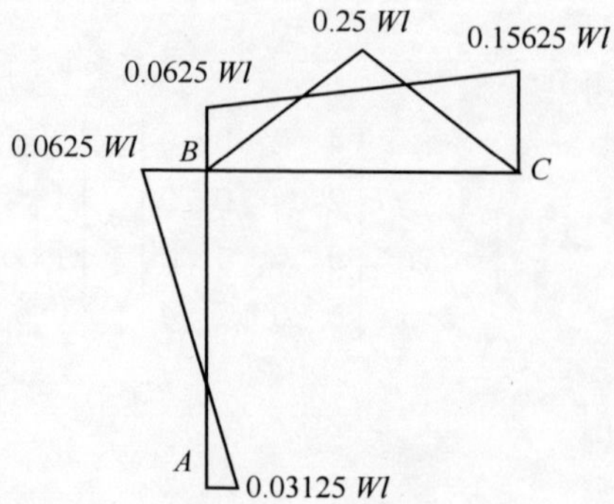

Final BMD

Fig. 3.8(e)

3.4 ANALYSIS OF PIN-JOINTED PLANE FRAMES

Pin-jointed plane frames are made up of slender members pin-connected at ends and are capable of taking loads at joints. The members of the pin-jointed plane frames are subiected to either tensile or compressive forces.

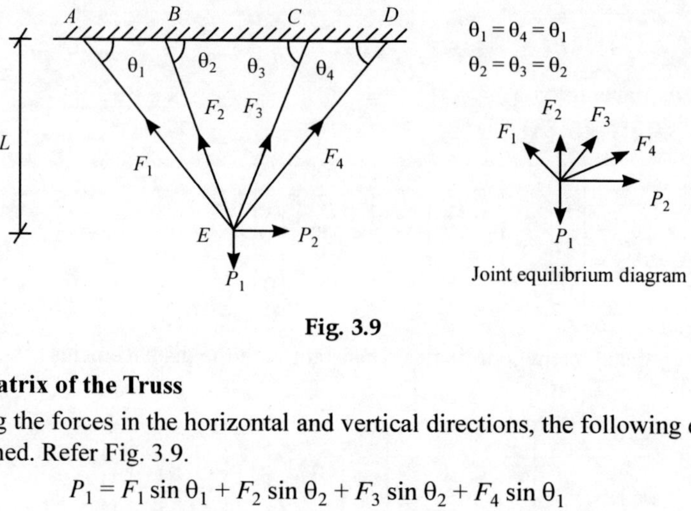

Fig. 3.9

Static Matrix of the Truss

Resolving the forces in the horizontal and vertical directions, the following equations are obtained. Refer Fig. 3.9.

$$P_1 = F_1 \sin \theta_1 + F_2 \sin \theta_2 + F_3 \sin \theta_2 + F_4 \sin \theta_1$$
$$P_2 = F_1 \cos \theta_1 + F_2 \cos \theta_2 - F_3 \cos \theta_2 - F_4 \cos \theta_1$$

or

$$\begin{Bmatrix} P_1 \\ P_2 \end{Bmatrix} = \begin{bmatrix} \sin\theta_1 & +\sin\theta_2 & +\sin\theta_2 & +\sin\theta_1 \\ \cos\theta_1 & +\cos\theta_2 & -\cos\theta_2 & -\cos\theta_1 \end{bmatrix} \times \begin{Bmatrix} F_1 \\ F_2 \\ F_3 \\ F_4 \end{Bmatrix}$$

$$\text{Statics matrix, } [A] = \begin{bmatrix} \sin\theta_1 & +\sin\theta_2 & +\sin\theta_2 & +\sin\theta_1 \\ \cos\theta_1 & +\cos\theta_2 & -\cos\theta_2 & -\cos\theta_1 \end{bmatrix}$$

$[\beta]$ Matrix $= [A]^T$

$$[\beta] = \begin{bmatrix} \sin\theta_1 & \cos\theta_1 \\ \sin\theta_2 & \cos\theta_2 \\ \sin\theta_2 & -\cos\theta_2 \\ \sin\theta_1 & -\cos\theta_1 \end{bmatrix}$$

Pin-jointed plane frames are analysed here by stiffness matrix method.

Example 3.9 Using matrix stiffness method, analyse the truss as shown in Fig. 3.10(a). *AE* and *L* are tabulated below for all the three members.

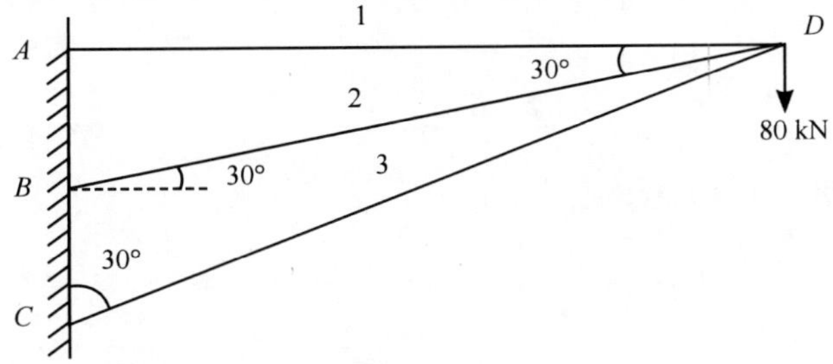

Fig. 3.10(a)

Member	*AE*	*L*
AD	400	400
BD	461.9	461.9
CD	800	800

Solution:

There are two possible displacements at *D*. Those are vertical and horizontal displacements.

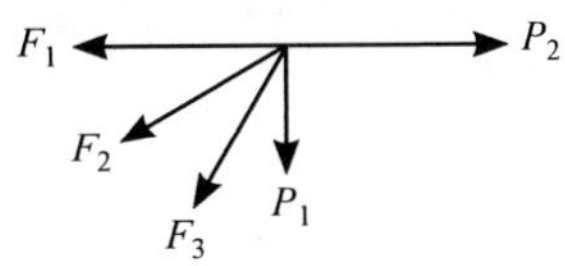

Joint equilibrium diagram

Fig. 3.10(b)

Static Matrix of the Truss

Resolving the forces in the vertical and horizontal directions, the following equations are obtained.

$$P_1 = F_1 \sin 0 - F_2 \sin 30 - F_3 \sin 60$$

$$P_2 = F_1 \cos 0 + F_2 \cos 30 + F_3 \cos 60$$

or

$$\begin{Bmatrix} P_1 \\ P_2 \end{Bmatrix} = \begin{bmatrix} \sin 0 & -\sin 30 & -\sin 60 \\ \cos 0 & \cos 30 & \cos 60 \end{bmatrix} \times \begin{Bmatrix} F_1 \\ F_2 \\ F_3 \end{Bmatrix}$$

Statics matrix, $$[A] = \begin{bmatrix} \sin 0 & -\sin 30 & -\sin 60 \\ \cos 0 & \cos 30 & \cos 60 \end{bmatrix}$$

$$[\beta] \text{ Matrix} = [A]^T$$

$$[\beta] = \begin{bmatrix} 0 & 1 \\ -0.5 & 0.87 \\ -0.87 & 0.5 \end{bmatrix}$$

Element Stiffness Matrix, $[k_n]$

$$[k_1] = \frac{AE}{L}[1]$$

Since $\frac{AE}{L} = 1$ for all members, $[k_1] = [k_2] = [k_3] = 1$

Assembled Stiffness Matrix, $[k]$

$$[k] = \begin{bmatrix} k_1 & 0 & 0 \\ 0 & k_2 & 0 \\ 0 & 0 & k_3 \end{bmatrix} = \begin{bmatrix} 1 & 0 & 0 \\ 0 & 1 & 0 \\ 0 & 0 & 1 \end{bmatrix}$$

System Stiffness Matrix, $[k]$

$$[k] = [\beta]^T[k][\beta]$$

$$[k] = \begin{bmatrix} 0 & -0.5 & -0.87 \\ 1 & 0.87 & 0.5 \end{bmatrix} \times \begin{bmatrix} 1 & 0 & 0 \\ 0 & 1 & 0 \\ 0 & 0 & 1 \end{bmatrix} \times \begin{bmatrix} 0 & 1 \\ -0.5 & 0.87 \\ -0.87 & 0.5 \end{bmatrix}$$

$$[k] = \begin{bmatrix} 1 & -0.87 \\ -0.87 & 2 \end{bmatrix}$$

$$[k]^{-1} = \frac{1}{1.25}\begin{bmatrix} 2 & 0.87 \\ 0.87 & 1 \end{bmatrix}$$

$$\{\Delta\} = [k]^{-1}[F]$$

$$\{\Delta\} = \frac{1}{1.25}\begin{bmatrix} 2 & 0.87 \\ 0.87 & 1 \end{bmatrix} \times \begin{Bmatrix} 80 \\ 0 \end{Bmatrix} = \begin{Bmatrix} 128 \\ 55.424 \end{Bmatrix}$$

Final Forces, $\{p^f\} = [k][\beta]\{\Delta\}$

$$\{p^f\} = \begin{bmatrix} 1 & 0 & 0 \\ 0 & 1 & 0 \\ 0 & 0 & 1 \end{bmatrix} \times \begin{bmatrix} 0 & 1 \\ -0.5 & 0.87 \\ -0.87 & 0.5 \end{bmatrix} \times \begin{Bmatrix} 128 \\ 55.424 \end{Bmatrix}$$

$$\{p^f\} = \begin{Bmatrix} 55.424 \\ -16.0 \\ -83.136 \end{Bmatrix}$$

Example 3.10 Using matrix stiffness method, analyse the truss as shown in Fig. 3.11(a). *AE* and *L* are tabulated below for all the three members.

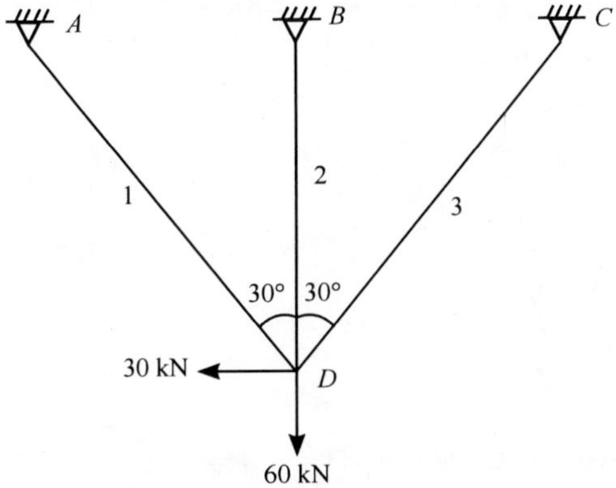

Fig. 3.11(a)

Member	*AE*	*L*
AD, CD	300	300
BD	259.8	259.8

Solution:

There are two possible displacements at *D*. Those are vertical and horizontal displacements.

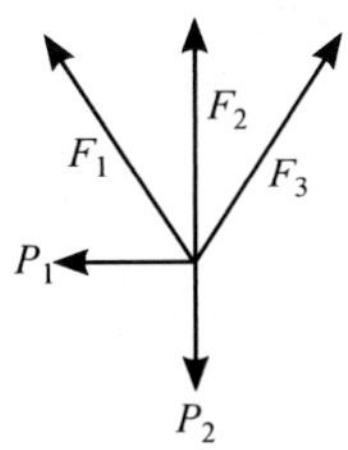

Joint equilibrium diagram

Fig. 3.11(b)

Static Matrix of the Truss

Resolving the forces in the horizontal and vertical directions, the following equations are obtained.

$$P_1 = -F_1 \cos 60 + F_2 \cos 90 + F_3 \cos 60$$
$$P_2 = F_1 \sin 60 + F_2 \sin 90 + F_3 \sin 60$$

or

$$\begin{Bmatrix} P_1 \\ P_2 \end{Bmatrix} = \begin{bmatrix} -\cos 60 & \cos 90 & \cos 60 \\ \sin 60 & \sin 90 & \sin 60 \end{bmatrix} \times \begin{Bmatrix} F_1 \\ F_2 \\ F_3 \end{Bmatrix}$$

Static matrix, $[A] = \begin{bmatrix} -\cos 60 & \cos 90 & \cos 60 \\ \sin 60 & \sin 90 & \sin 60 \end{bmatrix}$

$$[\beta] \text{ Matrix} = [A]^T$$

$$[\beta] = \begin{bmatrix} -0.5 & 0.87 \\ 0 & 1 \\ 0.5 & 0.87 \end{bmatrix}$$

Element Stiffness Matrix, $[k_n]$

$$[k_1] = \frac{AE}{L}[1]$$

Since $\frac{AE}{L} = 1$ for all members, $[k_1] = [k_2] = [k_3] = 1$

Assembled Stiffness Matrix, $[k]$

$$[k] = \begin{bmatrix} k_1 & 0 & 0 \\ 0 & k_2 & 0 \\ 0 & 0 & k_3 \end{bmatrix} = \begin{bmatrix} 1 & 0 & 0 \\ 0 & 1 & 0 \\ 0 & 0 & 1 \end{bmatrix}$$

System Stiffness Matrix, $[k]$

$$[k] = [\beta]^T[k][\beta]$$

$$[k] = \begin{bmatrix} -0.5 & 0 & 0.5 \\ 0.87 & 1 & 0.87 \end{bmatrix} \times \begin{bmatrix} 1 & 0 & 0 \\ 0 & 1 & 0 \\ 0 & 0 & 1 \end{bmatrix} \times \begin{bmatrix} -0.5 & 0.87 \\ 0 & 1 \\ 0.5 & 0.87 \end{bmatrix}$$

$$[k] = \begin{bmatrix} 0.5 & 0 \\ 0 & 2.5 \end{bmatrix}$$

$$[k]^{-1} = \frac{1}{1.25}\begin{bmatrix} 2.5 & 0 \\ 0 & 0.5 \end{bmatrix}$$

$$\{\Delta\} = [k]^{-1}[F]$$

$$\{\Delta\} = \frac{1}{1.25}\begin{bmatrix} 2.5 & 0 \\ 0 & 0.5 \end{bmatrix} \times \begin{Bmatrix} 30 \\ 60 \end{Bmatrix} = \begin{Bmatrix} 60 \\ 24 \end{Bmatrix}$$

Final Forces, $\{p^f\} = [k][\beta]\{\Delta\}$

$$\{p^f\} = \begin{bmatrix} 1 & 0 & 0 \\ 0 & 1 & 0 \\ 0 & 0 & 1 \end{bmatrix} \times \begin{bmatrix} -0.5 & 0.87 \\ 0 & 1 \\ 0.5 & 0.87 \end{bmatrix} \times \begin{Bmatrix} 60 \\ 24 \end{Bmatrix}$$

$$\{p^f\} = \begin{Bmatrix} -9.216 \\ 24 \\ 50.784 \end{Bmatrix}$$

3.5 BRIEF PROCEDURE FOR ANALYSIS OF INDETERMINATE STRUCTURES

In general, the following procedure is to be followed.

1. Determine the kinematic indeterminacy. This will form the system coordinates.
2. Split the structure ino elements and assign nodal coordinates for the elements. This forms element coordinates.
3. Evaluate element forces.
4. Form the transformation matrix [β].
5. From the element stiffness matrices, $[k_1]$, $[k_2]$,.....$[k_n]$ for all the elements in the structure, form $[k]$, the assembled element stiffness matrix.
6. Form the system stiffness matrix $[k]$ using $[k] = [\beta]^T[k][\beta]$
7. Get $[k]^{-1}$
8. Compute the system displacement vector $\{\Delta\}$ using the relation, $\{\Delta\} = [k]^{-1}[\{F'\} - \{F^0\}]$.
9. Get the element displacements, $\{\delta\} = [\beta]\{\Delta\}$.
10. Get the corresponding element forces $\{p'\}$, using $\{p'\} = [k][\delta]$.
11. Compute the final forces $\{p^f\} = \{p^0\} + \{p'\}$.
12. Finally sketch bending moment diagram.

3.6 ANALYSIS OF BEAMS USING DIRECT STIFFNESS METHOD

In the analysis of continuous beams, normally the axial deformation is negligible and hence only two unknowns may be taken at each end of the element.

Typical element and coordinates of displacements selected are as shown in Fig. 3.12(a).

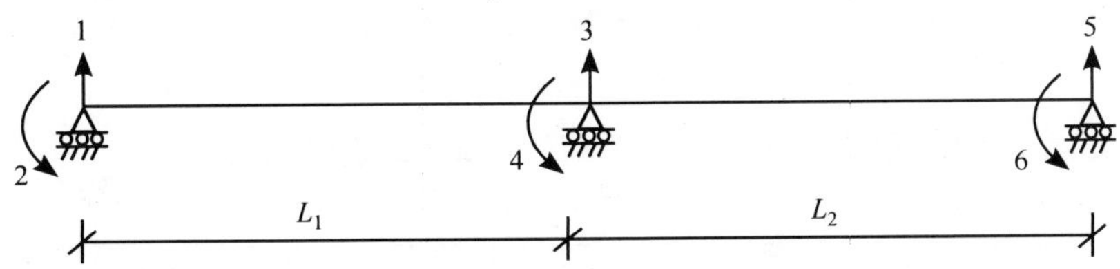

Fig. 3.12(a)

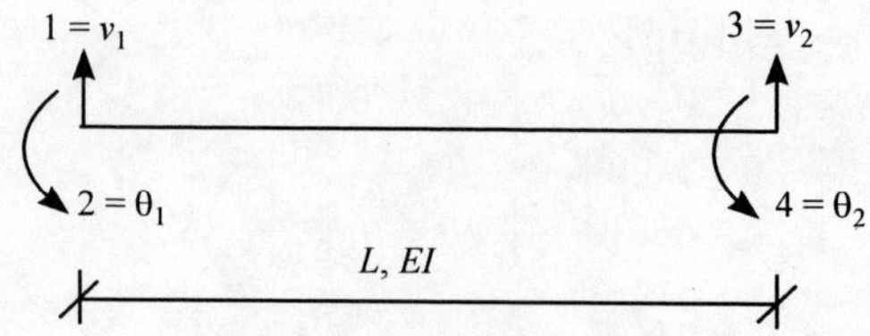

Fig. 3.12(b)

The end forces developed due to the unit displacement in all the four coordinate directions are as shown in Fig. 3.13.

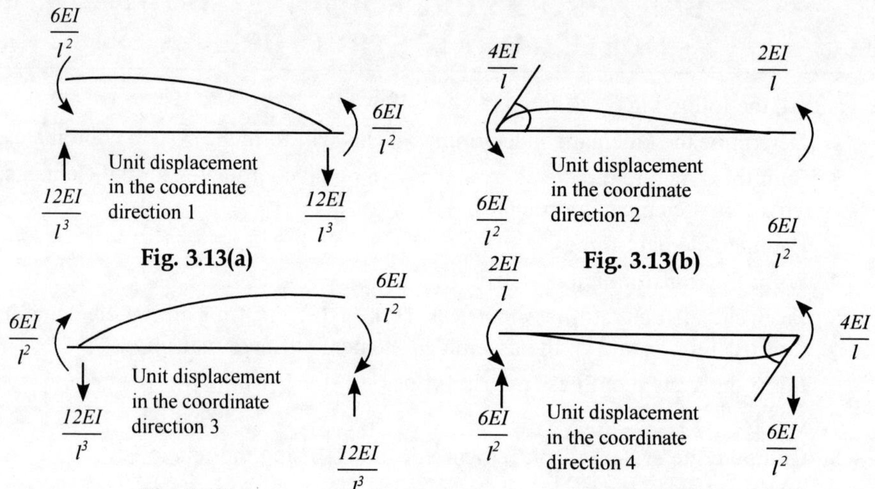

Fig. 3.13

$$[k] = \frac{EI}{l^3}\begin{bmatrix} 12 & 6L & -12 & 6L \\ 6L & 4L^2 & -6L & 2L^2 \\ -12 & -6L & 12 & -6L \\ 6L & 2L^2 & -6L & 4L^2 \end{bmatrix}$$

Example 3.11

Compute the slopes at the roller support points for the beam as shown in figure below. $E = 200$ GPa and $I = 4 \times 10^{-6}$ m^4. Refer Fig. 3.14(a).

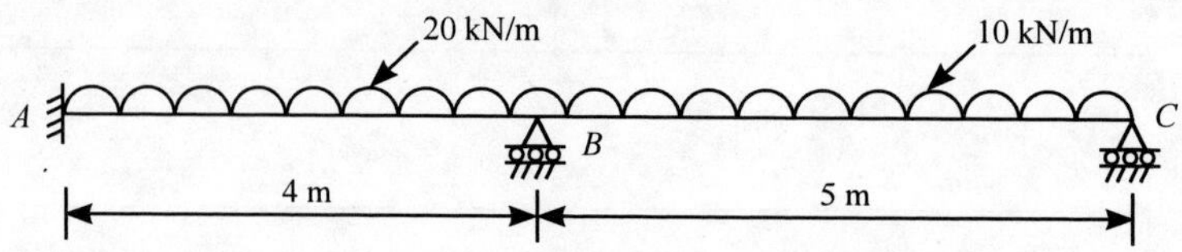

Fig. 3.14(a)

Solution:

$$E = 200 \text{ GPa} = 200 \times 10^9 \text{ Pa}$$
$$E = 200 \times 10^9 \text{ N/m}^2$$

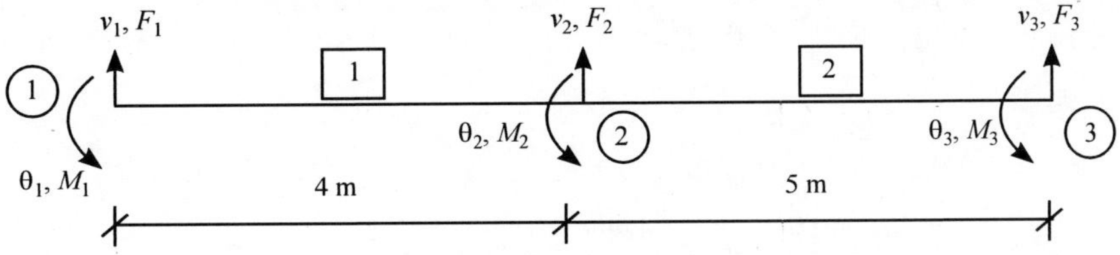

Fig. 3.14(b)

For element 1, at node 1, vertical deflection – v_1, slope deflection – θ_1, shear force – F_1 and bending moment – M_1

Similarly, for element 2, at node 2, vertical deflection – v_2, slope deflection – θ_2, shear force – F_2 and bending moment – M_2

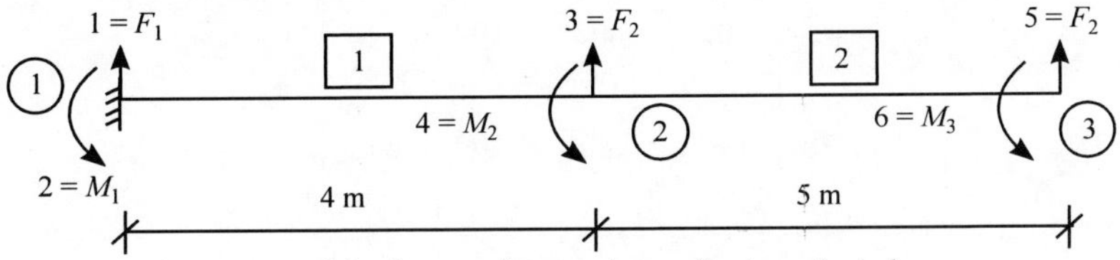

Member numbers and coordinates selected

Fig. 3.14(c)

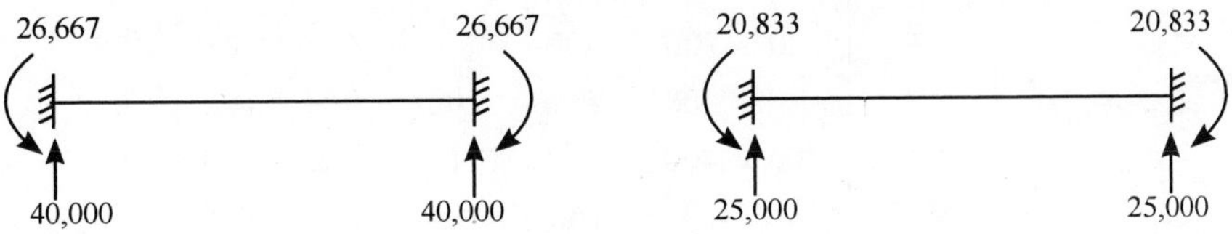

Fixed end forces

Fig. 3.14(d)

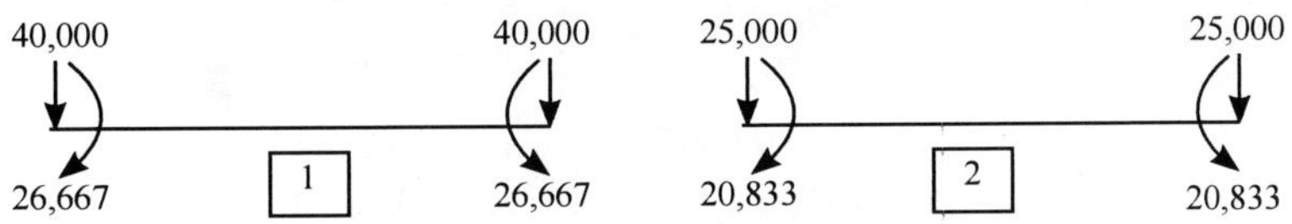

Equivalent load on joints

Fig. 3.14(e)

To determine the slopes at nodes 2 and 3.

The element stiffness matrix for element 1 and 2

$$[k_1] = \frac{E_1 I_1}{l_1^3} \begin{bmatrix} 12 & 6L & -12 & 6L \\ 6L & 4L^2 & -6L & 2L^2 \\ -12 & -6L & 12 & -6L \\ 6L & 2L^2 & -6L & 4L^2 \end{bmatrix}$$

$$\frac{E_1 I_1}{l_1^3} = 12.5 \times 10^3 \text{ N/m}$$

$$\frac{E_2 I_2}{l_2^3} = 6.4 \times 10^3 \text{ N/m}$$

$$[k_1] = 10^3 \begin{array}{c} \begin{matrix} v_1 & \theta_1 & v_2 & \theta_2 \end{matrix} \\ \begin{bmatrix} 150 & 300 & -150 & 300 \\ 300 & 800 & -300 & 400 \\ -150 & -300 & 150 & -300 \\ 300 & 400 & -300 & 800 \end{bmatrix} \end{array} \begin{matrix} v_1 \\ \theta_1 \\ v_2 \\ \theta_2 \end{matrix}$$

$$[k_2] = 10^3 \begin{array}{c} \begin{matrix} v_2 & \theta_2 & v_3 & \theta_3 \end{matrix} \\ \begin{bmatrix} 76.8 & 192 & -76.8 & 192 \\ 192 & 640 & -192 & 320 \\ -76.8 & -192 & 76.8 & -192 \\ 192 & 320 & -192 & 640 \end{bmatrix} \end{array} \begin{matrix} v_2 \\ \theta_2 \\ v_3 \\ \theta_3 \end{matrix}$$

The global stiffness matrix, $[k = k_1 + k_2]$

$$[k] = 10^3 \begin{array}{c} \begin{matrix} v_1 & \theta_1 & v_2 & \theta_2 & v_3 & \theta_3 \end{matrix} \\ \begin{bmatrix} 150 & 300 & -150 & 300 & 0 & 0 \\ 300 & 800 & -300 & 400 & 0 & 0 \\ -150 & 300 & 226.8 & -108 & -76.8 & 192 \\ 300 & 400 & -108 & 1440 & -192 & 320 \\ 0 & 0 & -76.8 & -192 & 76.8 & -192 \\ 0 & 0 & 192 & 320 & -192 & 640 \end{bmatrix} \end{array} \begin{matrix} v_1 \\ \theta_1 \\ v_2 \\ \theta_2 \\ v_3 \\ \theta_3 \end{matrix}$$

The force vector becomes

$$[F] = \begin{bmatrix} F_1 - 40{,}000 \\ M_1 - 26{,}667 \\ F_2 - 40{,}000 - 25{,}000 \\ M_2 + 26{,}667 - 20{,}833 \\ F_3 - 25{,}000 \\ M_3 + 20{,}833 \end{bmatrix} = \begin{bmatrix} F_1 - 40{,}000 \\ M_1 - 26{,}667 \\ F_2 - 65{,}000 \\ M_2 + 5834 \\ F_3 - 25{,}000 \\ M_3 + 20{,}833 \end{bmatrix}$$

Equation used: $[k]\{\Delta\} = \{F\}$

The boundary conditions for the given problem are:

At node 1, $v_1 = 0, \theta_1 = 0$

At node 2, $v_2 = 0, v_3 = 0$

There is no bending moment at nodes 2 and 3. Therefore, $M_2 = M_3 = 0$

$$[k]\{\Delta\} = \{F\}$$

$$[k] = 10^3 \begin{bmatrix} 150 & 300 & -150 & 300 & 0 & 0 \\ 300 & 800 & -300 & 400 & 0 & 0 \\ -150 & 300 & 226.8 & -108 & -76.8 & 192 \\ 300 & 400 & -108 & 1440 & -192 & 320 \\ 0 & 0 & -76.8 & -192 & 76.8 & -192 \\ 0 & 0 & 192 & 320 & -192 & 640 \end{bmatrix} \times \begin{Bmatrix} v_1 \\ \theta_1 \\ v_2 \\ \theta_2 \\ v_3 \\ \theta_3 \end{Bmatrix} = \begin{bmatrix} F_1 - 40,000 \\ M_1 - 26,667 \\ F_2 - 65,000 \\ M_2 + 5834 \\ F_3 - 25,000 \\ M_3 + 20,833 \end{bmatrix}$$

$$[k] = 10^3 \begin{bmatrix} 1440 & 320 \\ 320 & 640 \end{bmatrix} \begin{Bmatrix} \theta_2 \\ \theta_3 \end{Bmatrix} = \begin{bmatrix} 5834 \\ 20,833 \end{bmatrix}$$

$\theta_2 = -3.58 \times 10^{-3}$ rad

$\theta_3 = 34.3 \times 10^{-3}$ rad

3.7 ANALYSIS OF PIN-JOINTED PLANE FRAMES

Members of the truss are subjected to axial forces only but their orientation in the plane may be at an angle to the coordinate directions selected. Figure 3.15 shows a typical member of a truss with Young's modulus 'E', cross-sectional area 'A' and length 'L' at an angle 'θ'.

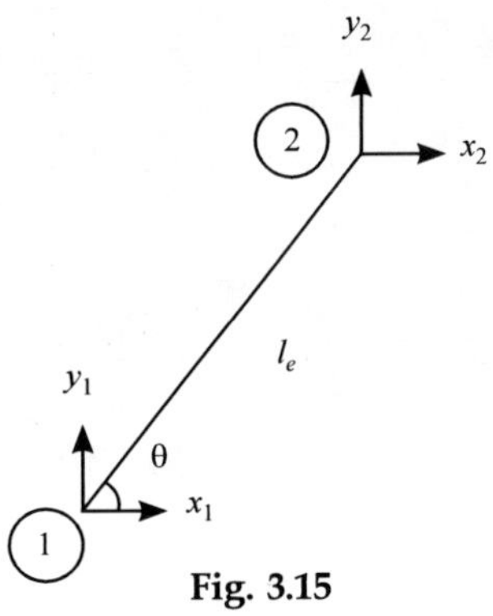

Fig. 3.15

If the coordinates (x_1, y_1) and (x_2, y_2) of the nodes 1 and 2 of the element are known, we can find,

$$l = \frac{x_2 - x_1}{l_e}$$

$$m = \frac{y_2 - y_1}{l_e}$$

where, l and m are the direction cosines.

$$l_e = \sqrt{(x_2 - x_1)^2 + (y_2 - y_1)^2}$$

The element stiffness matrix is

$$[k] = \frac{AE}{l}\begin{bmatrix} l^2 & lm & -l^2 & -lm \\ lm & m^2 & -lm & -m^2 \\ -l^2 & -lm & l^2 & lm \\ -lm & -m^2 & lm & m^2 \end{bmatrix}$$

Stress,
$$\sigma = \frac{E}{l}\begin{bmatrix} -l & -m & l & m \end{bmatrix}\begin{Bmatrix} \delta_1 \\ \delta_2 \\ \delta_3 \\ \delta_4 \end{Bmatrix}$$

Example 3.12 For a three-bar truss shown in below determine the nodal displacements and the stress in each member. Take E = 200 GPa. Refer Fig. 3.16(a).

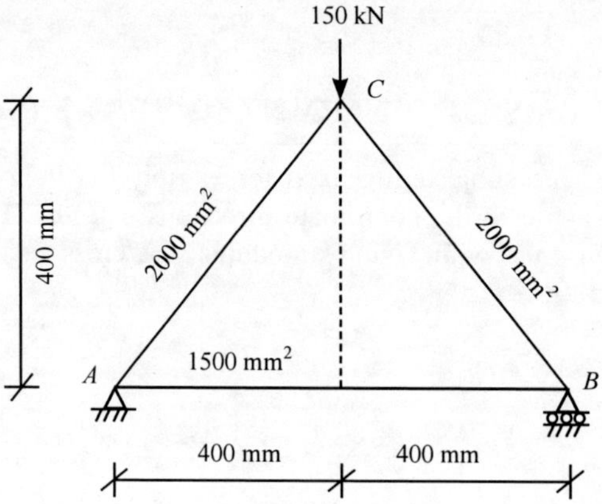

Fig. 3.16(a)

Solution:

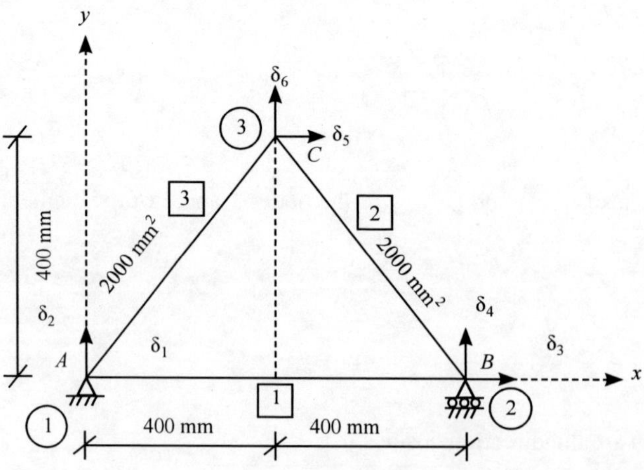

Fig. 3.16(b)

Taking node 1 as origin, the coordinates of various nodes are 1(0, 0), 2(800,0), 3(400,400)

$$l_{e1} = \sqrt{(800-0)^2 + (0-0)^2} \quad = 800 \text{ mm}$$

$$l_{e2} = \sqrt{(400-800)^2 + (400-0)^2} \quad = 565.69 \text{ mm}$$

$$l_{e3} = \sqrt{(400-0)^2 + (400-0)^2} \quad = 565.69 \text{ mm}$$

$$l_1 = \frac{x_2 - x_1}{le} = \frac{800-0}{800} = 1$$

$$m_1 = \frac{y_2 - y_1}{le} = \frac{0-0}{800} = 0$$

Similarly,

$$l_2 = -\,0.71$$
$$m_2 = 0.71$$
$$l_3 = 0.71$$
$$m_3 = 0.71$$

$$[k] = \frac{AE}{l}\begin{bmatrix} l^2 & lm & -l^2 & -lm \\ lm & m^2 & -lm & -m^2 \\ -l^2 & -lm & l^2 & lm \\ -lm & -m^2 & lm & m^2 \end{bmatrix}$$

$$[k]_{e1} = \frac{200\times 1500}{800}\begin{bmatrix} 1 & 0 & -1 & 0 \\ 0 & 0 & 0 & 0 \\ -1 & 0 & 1 & 0 \\ 0 & 0 & 0 & 0 \end{bmatrix}$$

$$\begin{matrix} \quad\ \ \delta_1 & \delta_2 & \delta_3 & \delta_4 \end{matrix}$$

$$[k]_{e1} = \begin{bmatrix} 375 & 0 & -375 & 0 \\ 0 & 0 & 0 & 0 \\ -375 & 0 & 375 & 0 \\ 0 & 0 & 0 & 0 \end{bmatrix} \begin{matrix} \delta_1 \\ \delta_2 \\ \delta_3 \\ \delta_4 \end{matrix}$$

Similarly,

$$\begin{matrix} \quad\ \ \delta_3 & \delta_4 & \delta_5 & \delta_6 \end{matrix}$$

$$[k]_{e2} = \begin{bmatrix} 353.55 & -353.55 & -353.55 & 353.55 \\ -353.55 & 353.55 & 353.55 & -353.55 \\ -353.55 & 353.55 & 353.55 & -353.55 \\ 353.55 & -353.55 & -353.55 & 353.55 \end{bmatrix} \begin{matrix} \delta_3 \\ \delta_4 \\ \delta_5 \\ \delta_6 \end{matrix}$$

Similarly,

$$[k]_{e3} = \begin{array}{c} \begin{matrix} \delta_1 & \quad\quad\delta_2 & \quad\quad\delta_5 & \quad\quad\delta_6 \end{matrix} \\ \begin{bmatrix} 353.55 & 353.55 & -353.55 & -353.55 \\ 353.55 & 353.55 & -353.55 & -353.55 \\ -353.55 & -353.55 & 353.55 & 353.55 \\ -353.55 & -353.55 & 353.55 & 353.55 \end{bmatrix} \end{array} \begin{matrix} \\ \delta_1 \\ \delta_2 \\ \delta_5 \\ \delta_6 \end{matrix}$$

$$[k] = [k]_{e1} + [k]_{e2} + [k]_{e3}$$

$$[k] = \begin{array}{c} \begin{matrix} \delta_1 & \quad\quad\delta_2 & \quad\quad\delta_3 & \quad\quad\delta_4 & \quad\quad\delta_5 & \quad\quad\delta_6 \end{matrix} \\ \begin{bmatrix} 728.55 & 353.55 & -375 & 0 & -353.55 & -353.55 \\ 353.55 & 353.55 & 0 & 0 & -353.55 & -353.55 \\ -375 & 0 & 728.55 & -353.55 & -353.55 & 353.55 \\ 0 & 0 & -353.55 & 353.55 & 353.55 & -353.55 \\ -353.55 & -353.55 & -353.55 & 353.55 & 707.1 & 0 \\ -353.55 & 353.55 & 353.55 & -353.55 & 0 & 707.1 \end{bmatrix} \end{array} \begin{matrix} \\ \delta_1 \\ \delta_2 \\ \delta_3 \\ \delta_4 \\ \delta_5 \\ \delta_6 \end{matrix}$$

Equation used: $[\boldsymbol{k}]\{\Delta\} = \{\boldsymbol{F}\}$

$$\begin{bmatrix} 728.55 & 353.55 & -375 & 0 & -353.55 & -353.55 \\ 353.55 & 353.55 & 0 & 0 & -353.55 & -353.55 \\ -375 & 0 & 728.55 & -353.55 & -353.55 & 353.55 \\ 0 & 0 & -353.55 & 353.55 & 353.55 & -353.55 \\ -353.55 & -353.55 & -353.55 & 353.55 & 707.1 & 0 \\ -353.55 & 353.55 & 353.55 & -353.55 & 0 & 707.1 \end{bmatrix} \times \begin{Bmatrix} \delta_1 \\ \delta_2 \\ \delta_3 \\ \delta_4 \\ \delta_5 \\ \delta_6 \end{Bmatrix} = \begin{Bmatrix} F_1 \\ F_2 \\ F_3 \\ F_4 \\ F_5 \\ F_6 \end{Bmatrix}$$

$$F_1 = F_2 = F_3 = F_4 = F_5 = 0 \text{ and } F_6 = -150$$

The boundary conditions are:

$$\delta_1 = \delta_2 = \delta_4 = 0$$

Hence, the equation reduces to

$$\begin{bmatrix} 728.55 & -353.55 & 353.55 \\ -353.55 & 707.1 & 0 \\ 353.55 & 0 & 707.1 \end{bmatrix} \times \begin{Bmatrix} \delta_3 \\ \delta_5 \\ \delta_6 \end{Bmatrix} = \begin{Bmatrix} 0 \\ 0 \\ -150 \end{Bmatrix}$$

$$\delta_3 = 0.2 \text{ mm}$$

$$\delta_5 = 0.1 \text{ mm}$$

$$\delta_6 = -\,0.312 \text{ mm}$$

$$\sigma = \frac{E}{l}\,[-l \quad -m \quad l \quad m] \begin{Bmatrix} \delta_1 \\ \delta_2 \\ \delta_3 \\ \delta_4 \end{Bmatrix}$$

$$\sigma_1 = 0.05 \text{ kN/mm}^2$$
$$\sigma_2 = -0.053 \text{ kN/mm}^2$$
$$\sigma_3 = -0.053 \text{ kN/mm}^2$$

Example 3.13

For a two-bar truss shown in below determine the displacement at node 2 and the stress in both the members. Take $E = 70$ GPa and $A = 200$ mm^2. Refer Fig. 3.17(a).

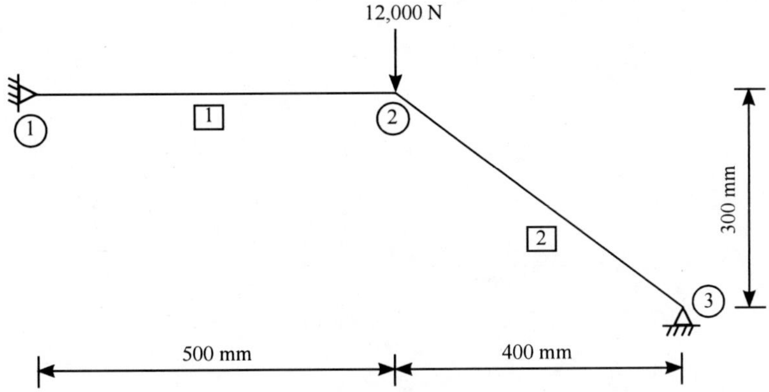

Fig. 3.17(a)

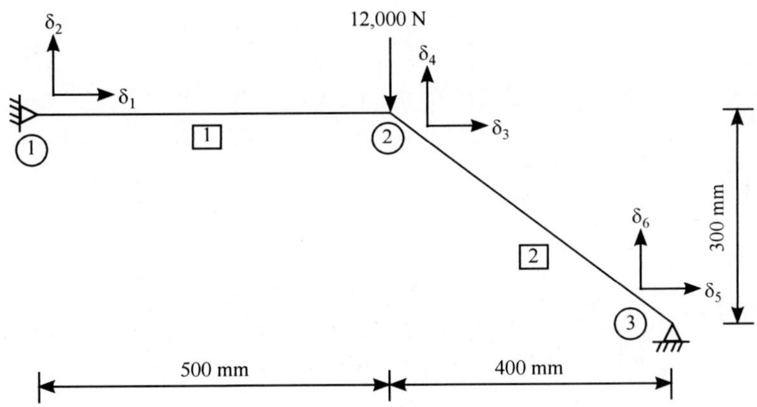

Fig. 3.17(b)

Solution:

Taking node 1 as origin, the coordinates of various nodes are 1(0, 0), 2(500,0), 3(900, –300)

$$l_{e1} = \sqrt{(500-0)^2 + (0-0)^2} = 500 \text{ mm}$$

$$l_{e2} = \sqrt{(900-500)^2 + (-300-0)^2} = 500 \text{ mm}$$

$$l_1 = \frac{x_2 - x_1}{l_e} = \frac{500-0}{500} = 1$$

$$m_1 = \frac{y_2 - y_1}{l_e} = \frac{0-0}{500} = 0$$

Similarly,

$$l_2 = 0.8$$
$$m_2 = -0.6$$

$$[k] = \frac{AE}{l}\begin{bmatrix} l^2 & lm & -l^2 & -lm \\ lm & m^2 & -lm & -m^2 \\ -l^2 & -lm & l^2 & lm \\ -lm & -m^2 & lm & m^2 \end{bmatrix}$$

$$[k]_{e1} = \frac{70\times 10^3 \times 200}{500}\begin{bmatrix} 1 & 0 & -1 & 0 \\ 0 & 0 & 0 & 0 \\ -1 & 0 & 1 & 0 \\ 0 & 0 & 0 & 0 \end{bmatrix}$$

$$\begin{matrix} & \delta_1 & \delta_2 & \delta_3 & \delta_4 & \\ [k]_{e1} = 28\times 10^3 & \begin{bmatrix} 1 & 0 & -1 & 0 \\ 0 & 0 & 0 & 0 \\ -1 & 0 & 1 & 0 \\ 0 & 0 & 0 & 0 \end{bmatrix} & & & & \begin{matrix} \delta_1 \\ \delta_2 \\ \delta_3 \\ \delta_4 \end{matrix} \end{matrix}$$

Similarly,

$$\begin{matrix} & \delta_3 & \delta_4 & \delta_5 & \delta_6 & \\ [k]_{e2} = 28\times 10^3 & \begin{bmatrix} 0.64 & -0.48 & -0.64 & 0.48 \\ -0.48 & 0.36 & 0.48 & -0.36 \\ -0.64 & 0.48 & 0.64 & -0.48 \\ 0.48 & -0.36 & -0.48 & 0.36 \end{bmatrix} & & & & \begin{matrix} \delta_3 \\ \delta_4 \\ \delta_5 \\ \delta_6 \end{matrix} \end{matrix}$$

Global stiffness matrix, $[k] = [k]_{e1} + [k]_{e2}$

$$\begin{matrix} & \delta_1 & \delta_2 & \delta_3 & \delta_4 & \delta_5 & \delta_6 & \\ [k] = 28\times 10^3 & \begin{bmatrix} 1 & 0 & -1 & 0 & 0 & 0 \\ 0 & 0 & 0 & 0 & 0 & 0 \\ -1 & 0 & 1.64 & -0.48 & -0.64 & 0.48 \\ 0 & 0 & -0.48 & 0.36 & 0.48 & -0.36 \\ 0 & 0 & -0.64 & 0.48 & 0.64 & -0.48 \\ 0 & 0 & 0.48 & -0.36 & -0.48 & 0.36 \end{bmatrix} & & & & & & \begin{matrix} \delta_1 \\ \delta_2 \\ \delta_3 \\ \delta_4 \\ \delta_5 \\ \delta_6 \end{matrix} \end{matrix}$$

Equation used: $[k]\{\Delta\} = \{F\}$

$$28 \times 10^3 \begin{bmatrix} 1 & 0 & -1 & 0 & 0 & 0 \\ 0 & 0 & 0 & 0 & 0 & 0 \\ -1 & 0 & 1.64 & -0.48 & -0.64 & 0.48 \\ 0 & 0 & -0.48 & 0.36 & 0.48 & -0.36 \\ 0 & 0 & -0.64 & 0.48 & 0.64 & -0.48 \\ 0 & 0 & 0.48 & -0.36 & -0.48 & 0.36 \end{bmatrix} \times \begin{Bmatrix} \delta_1 \\ \delta_2 \\ \delta_3 \\ \delta_4 \\ \delta_5 \\ \delta_6 \end{Bmatrix} = \begin{Bmatrix} F_1 \\ F_2 \\ F_3 \\ F_4 \\ F_5 \\ F_6 \end{Bmatrix}$$

$$F_1 = F_2 = F_3 = F_5 = F_6 = 0 \text{ and } F_4 = -12{,}000 \text{ N}$$

The boundary conditions are:

$$\delta_1 = \delta_2 = 0$$
$$\delta_5 = \delta_6 = 0$$

Hence, the equation reduces to

$$28 \begin{bmatrix} 1.64 & -0.48 \\ -0.48 & 0.36 \end{bmatrix} \times \begin{Bmatrix} \delta_3 \\ \delta_4 \end{Bmatrix} = \begin{Bmatrix} 0 \\ -12 \end{Bmatrix}$$

$$\delta_3 = -0.57 \text{ mm}$$
$$\delta_4 = -1.96 \text{ mm}$$

The stress in the member is given by

$$\sigma = \frac{E}{l}\begin{bmatrix} -l & -m & l & m \end{bmatrix} \begin{Bmatrix} \delta_1 \\ \delta_2 \\ \delta_3 \\ \delta_4 \end{Bmatrix}$$

$$\sigma_1 = -79.8 \text{ N/mm}^2$$

3.8 ANALYSIS OF PLANE FRAMES USING DIRECT STIFFNESS METHOD

Frame element is a combination of bar element and also a beam element as shown in Fig. 3.18.

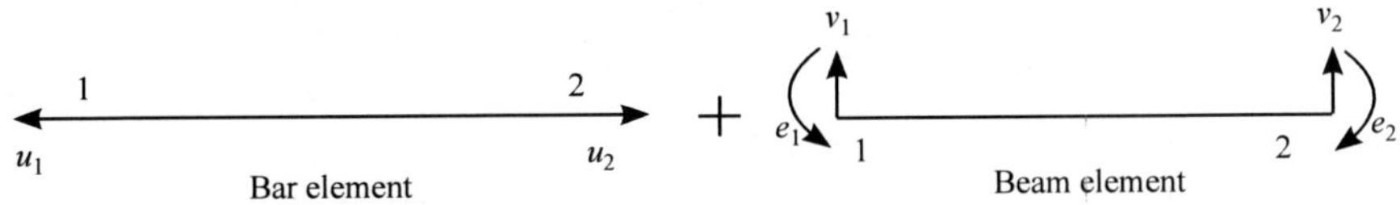

Fig. 3.18

In case of bar element, displacement will be in the axial direction. In case of beam element, two types of displacements are seen, one is transverse displacement and the other is rotational displacement.

Stiffness matrix of a bar element is given by

$$[k]_{\text{bar}} \text{ element} = \frac{AE}{L}\begin{bmatrix} 1 & -1 \\ -1 & 1 \end{bmatrix} = \begin{bmatrix} \frac{AE}{L} & -\frac{AE}{L} \\ -\frac{AE}{L} & \frac{AE}{L} \end{bmatrix} \quad ...(1)$$

Similarly, stiffness matrix for the beam element is given by

$$[k]_{\text{beam}} \text{ element} = \frac{EI}{L^3}\begin{bmatrix} 12 & 6L & -12 & 6L \\ 6L & 4L^2 & -6L & 2L^2 \\ -12 & -6L & 12 & -6L \\ 6L & 2L^2 & -6L & 4L^2 \end{bmatrix} \quad ...(2)$$

The combined stiffness matrix for the frame element is given by

$[k]_{\text{Frame element}} = [k]_{\text{bar element}} + [k]_{\text{beam element}}$

Let, $G = \dfrac{AE}{L}$ and $H = \dfrac{EI}{L^3}$

Then equation (1) becomes

$$[k]_{\text{bar element}} = \begin{array}{c} \begin{matrix} u_1 & u_2 \end{matrix} \\ \begin{bmatrix} G & -G \\ -G & G \end{bmatrix} \end{array} \begin{matrix} u_1 \\ u_2 \end{matrix} \quad ...(3)$$

The equation (2) becomes

$$[k]_{\text{beam element}} = \begin{array}{c} \begin{matrix} v_1 & \theta_1 & v_2 & \theta_2 \end{matrix} \\ \begin{bmatrix} 12H & 6HL & -12H & 6HL \\ 6HL & 4HL^2 & -6HL & 2HL^2 \\ -12H & -6HL & 12H & -6HL \\ 6HL & 2HL^2 & -6HL & 4HL^2 \end{bmatrix} \end{array} \begin{matrix} v_1 \\ \theta_1 \\ v_2 \\ \theta_2 \end{matrix} \quad ...(4)$$

$$[k]_{\text{Frame element}} = \begin{array}{c} \begin{matrix} u_1 & v_1 & \theta_1 & u_2 & v_2 & \theta_2 \end{matrix} \\ \begin{bmatrix} G & 0 & 0 & -G & 0 & 0 \\ 0 & 12H & 6HL & 0 & 12H & 6HL \\ 0 & 6HL & 4HL^2 & 0 & -6HL & -HL^2 \\ -G & 0 & 0 & G & 0 & 0 \\ 0 & -12H & -6H & 0 & 12H & -6HL \\ 0 & 6HL & 2HL^2 & 0 & -6HL & 4HL^2 \end{bmatrix} \end{array} \begin{matrix} u_1 \\ v_1 \\ \theta_1 \\ u_2 \\ v_2 \\ \theta_2 \end{matrix}$$

Example 3.14 Find the unknown deformations at the nodes 1 and 2 of the frame as shown in Fig. 3.19(a). Use $E = 200$ GPa, $A = 0.0112$ m^2 and $I = 2.39 \times 10^{-5}$ m^2.

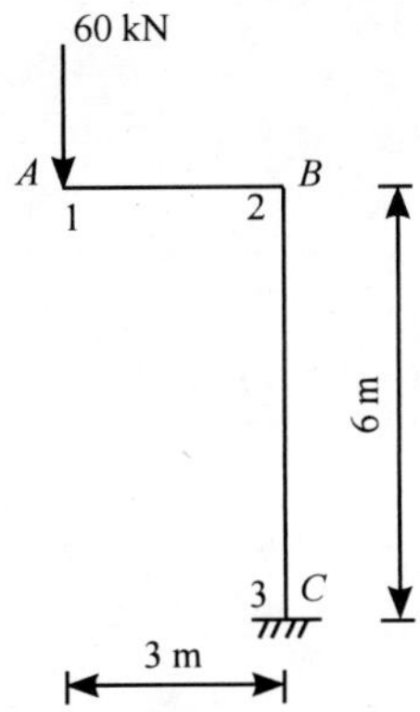

Fig. 3.19(a)

Solution:

Mark displacements at each node as shown in Fig. 3.19(b).

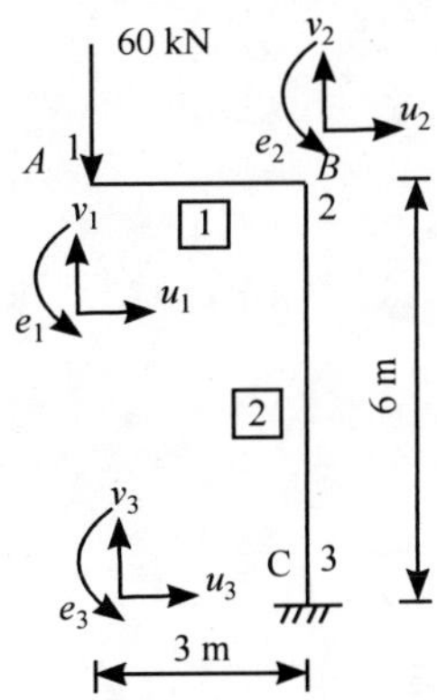

Fig. 3.19(b)

Taking node 1 as origin, the coordinates of various nodes are 1(0, 0), 2(3,0), 3(3, -6)

The stiffness matrix for element 1

Length of the element is given by

$$L_e = \sqrt{(x_2 - x_1)^2 + (y_2 - y_1)^2}$$

$$L_e = \sqrt{(3-0)^2 + (0-0)^2} = 3 \text{ m}$$

$$c = \frac{x_2 - x_1}{L} = \frac{3-0}{3} = 1$$

$$s = \frac{y_2 - y_1}{L} = \frac{0-0}{3} = 0$$

Stiffness Matrix, $[k]$ =

$$\frac{E}{L}\begin{bmatrix} \left(Ac^2+\frac{12I}{L^2}s^2\right) & \left(A-\frac{12I}{L^2}\right)cs & -\frac{6I}{L}s & -\left(Ac^2+\frac{12I}{L^2}s^2\right) & -\left(A-\frac{12I}{L^2}\right)cs & -\frac{6I}{L} \\ \left(A-\frac{12I}{L^2}\right)cs & \left(As^2+\frac{12I}{L^2}c^2\right) & \frac{6I}{L}c & -\left(A-\frac{12I}{L^2}\right)cs & -\left(As^2+\frac{12I}{L^2}c^2\right) & \frac{6I}{L}c \\ -\frac{6I}{L}s & \frac{6I}{L}c & 4I & \frac{6I}{L}s & -\frac{6I}{L}c & 2I \\ -\left(Ac^2+\frac{12I}{L^2}s^2\right) & -\left(A-\frac{12I}{L^2}\right)cs & \frac{6I}{L}s & \left(Ac^2+\frac{12I}{L^2}s^2\right) & \left(A-\frac{12I}{L^2}\right)cs & \frac{6I}{L}s \\ -\left(A-\frac{12I}{L^2}\right)cs & -\left(As^2+\frac{12I}{L^2}c^2\right) & -\frac{6I}{L}c & \left(A-\frac{12I}{L^2}\right)cs & \left(As^2+\frac{12I}{L^2}c^2\right) & -\frac{6I}{L} \\ -\frac{6I}{L}s & \frac{6I}{L}c & 2I & \frac{6I}{L}s & -\frac{6I}{L}c & 4I \end{bmatrix}$$

$$\begin{matrix} & u_1 & v_1 & \theta_1 & u_2 & v_2 & \theta_2 & \end{matrix}$$

$$[k]_{e1} = 6.67\times10^{10}\begin{bmatrix} 0.0112 & 0 & 0 & -0.0112 & 0 & 0 \\ 0 & 3.19\times10^{-5} & 4.7\times10^{-5} & 0 & -3.19\times10^{-5} & 4.7\times10^{-5} \\ 0 & 4.7\times10^{-5} & 9.5\times10^{-5} & 0 & -4.78\times10^{-5} & 4.7\times10^{-5} \\ -0.0112 & 0 & 0 & 0.0112 & 0 & 0 \\ 0 & -3.19\times10^{-5} & 4.7\times10^{-5} & 0 & 3.19\times10^{-5} & 4.78\times10^{-5} \\ 0 & -4.7\times10^{-5} & 4.78\times10^{-5} & 0 & 4.78\times10^{-5} & 9.56\times10^{-5} \end{bmatrix}\begin{matrix} u_1 \\ v_1 \\ \theta_1 \\ u_2 \\ v_2 \\ \theta_2 \end{matrix}$$

The stiffness matrix for element 2

Length of the element is given by

$$L = \sqrt{(x_3-x_2)^2+(y_3-y_2)^2}$$

$$L = \sqrt{(3-3)^2+(-6-0)^2} = 6 \text{ m}$$

$$c = \frac{x_3-x_2}{L} = \frac{3-3}{6} = 0$$

$$s = \frac{y_3-y_2}{L} = \frac{-6-0}{6} = -1$$

$$\begin{matrix} & u_2 & v_2 & \theta_2 & u_3 & v_3 & \theta_3 & \end{matrix}$$

$$[k]_{e2} = 3.33\times10^{10}\begin{bmatrix} 7.79\times10^{-6} & 0 & 2.39\times10^{-5} & -7.7\times10^{-6} & 0 & 2.3\times10^{-5} \\ 0 & 0.0112 & 0 & 0 & -0.0112 & 0 \\ 2.39\times10^{-5} & 0 & 9.5\times10^{-5} & -2.39\times10^{-5} & 0 & 4.78\times10^{-5} \\ -7.9\times10^{-1} & 0 & 2.39\times10^{-5} & 7.7\times10^{-6} & 0 & 2.3\times10^{-5} \\ 0 & -0.0112 & 0 & 0 & 0.0012 & 0 \\ 2.39\times10^{-5} & 0 & 4.78\times10^{-5} & 2.39\times10^{-5} & 0 & 9.56\times10^{-5} \end{bmatrix}\begin{matrix} u_2 \\ v_2 \\ \theta_2 \\ u_3 \\ v_3 \\ \theta_3 \end{matrix}$$

$$[k] = [k]_{e1} + [k]_{e2}$$

$$
[k] = 1\times 10^{11}
\begin{array}{c}
\begin{array}{ccccccccc}
u_1 & v_1 & \theta_1 & u_2 & v_2 & \theta_2 & u_3 & v_3 & \theta_3
\end{array} \\
\begin{bmatrix}
0.0112 & 0 & 0 & -0.0112 & 0 & 0 & 0 & 0 & 0 \\
0 & 3.19\times10^{-5} & 4.7\times10^{-5} & 0 & -3.19\times10^{-5} & 4.7\times10^{-5} & 0 & 0 & 0 \\
0 & 4.7\times10^{-5} & 9.5\times10^{-5} & 0 & -4.78\times10^{-5} & 4.7\times10^{-5} & 0 & 0 & 0 \\
-0.0112 & 0 & 0 & 11.21\times10^{-3} & 0 & 2.39\times10^{-5} & -7.7\times10^{-6} & 0 & 2.3\times10^{-5} \\
0 & -3.19\times10^{-5} & 4.7\times10^{-5} & 0 & 3.19\times10^{-5} & 4.78\times10^{-5} & 0 & -0.112 & 0 \\
0 & -4.7\times10^{-5} & 4.78\times10^{-5} & 2.39\times10^{-5} & 4.78\times10^{-5} & 1.9\times10^{-4} & -2.39\times10^{-5} & 0 & 4.78\times10^{-5} \\
0 & 0 & 0 & -7.9\times10^{-1} & 0 & 2.39\times10^{-5} & 7.7\times10^{-6} & 0 & 2.3\times10^{-5} \\
0 & 0 & 0 & 0 & -0.0112 & 0 & 0 & 0.0012 & 0 \\
0 & 0 & 0 & -2.39\times10^{-5} & 0 & 4.78\times10^{-5} & 2.39 & 0 & 9.56\times10^{-5}
\end{bmatrix}
\end{array}
\begin{array}{c} u_1 \\ v_1 \\ \theta_1 \\ u_2 \\ v_2 \\ \theta_2 \\ u_3 \\ v_3 \\ \theta_3 \end{array}
$$

Equation used: $[k]\{\Delta\} = \{F\}$

$$
[k] = 1\times 10^{11}
\begin{bmatrix}
0.0112 & 0 & 0 & -0.0112 & 0 & 0 & 0 & 0 & 0 \\
0 & 3.19\times10^{-5} & 4.7\times10^{-5} & 0 & -3.19\times10^{-5} & 4.7\times10^{-5} & 0 & 0 & 0 \\
0 & 4.7\times10^{-5} & 9.5\times10^{-5} & 0 & -4.78\times10^{-5} & 4.7\times10^{-5} & 0 & 0 & 0 \\
-0.0112 & 0 & 0 & 11.21\times10^{-3} & 0 & 2.39\times10^{-5} & -7.7\times10^{-6} & 0 & 2.3\times10^{-5} \\
0 & -3.19\times10^{-5} & 4.7\times10^{-5} & 0 & 3.19\times10^{-5} & 4.78\times10^{-5} & 0 & -0.112 & 0 \\
0 & -4.7\times10^{-5} & 4.78\times10^{-5} & 2.39\times10^{-5} & 4.78\times10^{-5} & 1.9\times10^{-4} & -2.39\times10^{-5} & 0 & 4.78\times10^{-5} \\
0 & 0 & 0 & -7.9\times10^{-1} & 0 & 2.39\times10^{-5} & 7.7\times10^{-6} & 0 & 2.3\times10^{-5} \\
0 & 0 & 0 & 0 & -0.0112 & 0 & 0 & 0.0012 & 0 \\
0 & 0 & 0 & -2.39\times10^{-5} & 0 & 4.78\times10^{-5} & 2.39 & 0 & 9.56\times10^{-5}
\end{bmatrix}
\times
\begin{Bmatrix} u_1 \\ v_1 \\ \theta_1 \\ u_2 \\ v_2 \\ \theta_2 \\ u_3 \\ v_3 \\ \theta_3 \end{Bmatrix}
=
\begin{Bmatrix} F_{1x} \\ F_{1y} \\ M_1 \\ F_{2x} \\ F_{2y} \\ M_2 \\ F_{3x} \\ F_{3y} \\ M_3 \end{Bmatrix}
$$

Boundary Conditions

$$u_3 = v_3 = \theta_3 = 0$$

$$F_{1y} = -60{,}000 \text{ N}$$

Unknowns to be determined

u_1, v_1, θ_1 and u_2, v_2, θ_2

$$
[k] = 1\times 10^{11}
\left[
\begin{array}{cccccc|ccc}
\multicolumn{6}{c|}{k_{11}} & \multicolumn{3}{c}{k_{12}} \\
0.0112 & 0 & 0 & -0.0112 & 0 & 0 & 0 & 0 & 0 \\
0 & 3.19\times10^{-5} & 4.7\times10^{-5} & 0 & -3.19\times10^{-5} & 4.7\times10^{-5} & 0 & 0 & 0 \\
0 & 4.7\times10^{-5} & 9.5\times10^{-5} & 0 & -4.78\times10^{-5} & 4.7\times10^{-5} & 0 & 0 & 0 \\
-0.0112 & 0 & 0 & 11.21\times10^{-3} & 0 & 2.39\times10^{-5} & -7.7\times10^{-6} & 0 & 2.3\times10^{-5} \\
0 & -3.19\times10^{-5} & 4.7\times10^{-5} & 0 & 3.19\times10^{-5} & 4.78\times10^{-5} & 0 & -0.112 & 0 \\
0 & -4.7\times10^{-5} & 4.78\times10^{-5} & 2.39\times10^{-5} & 4.78\times10^{-5} & 1.9\times10^{-4} & -2.39\times10^{-5} & 0 & 4.78\times10^{-5} \\
\hline
0 & 0 & 0 & -7.9\times10^{-1} & 0 & 2.39\times10^{-5} & 7.7\times10^{-6} & 0 & 2.3\times10^{-5} \\
0 & 0 & 0 & 0 & -0.0112 & 0 & 0 & 0.0012 & 0 \\
0 & 0 & 0 & -2.39\times10^{-5} & 0 & 4.78\times10^{-5} & 2.39 & 0 & 9.56\times10^{-5} \\
\multicolumn{6}{c|}{k_{21}} & \multicolumn{3}{c}{k_{22}}
\end{array}
\right]
\times
\begin{Bmatrix} u_1 \\ v_1 \\ \theta_1 \\ u_2 \\ v_2 \\ \theta_2 \\ \hline 0 \\ 0 \\ 0 \end{Bmatrix}
=
\begin{Bmatrix} 0 \\ -60000 \\ 0 \\ 0 \\ 0 \\ 0 \\ \hline F_{3x} \\ F_{3y} \\ M_3 \end{Bmatrix}
$$

$$[k_{11}]\{\Delta_u\} + [k_{12}]\{\Delta_k\} = \{F_k\}$$

$$[k_{21}]\{\Delta_u\} + [k_{22}]\{\Delta_k\} = \{F_u\}$$

$$\{\Delta_u\} = \begin{Bmatrix} u_1 \\ v_1 \\ \theta_1 \\ u_2 \\ v_2 \\ \theta_2 \end{Bmatrix}$$

$$\{\Delta_k\} = \begin{Bmatrix} 0 \\ 0 \\ 0 \end{Bmatrix}$$

$$\{F_k\} = \begin{Bmatrix} 0 \\ -60,000 \\ 0 \\ 0 \\ 0 \\ 0 \end{Bmatrix}$$

$$\{F_u\} = \begin{Bmatrix} F_{3x} \\ F_{3y} \\ M_3 \end{Bmatrix}$$

$$[k_{11}]\{\Delta_u\} + [k_{12}]\{\Delta_k\} = \{F_k\}$$

$$\begin{bmatrix} 1.12\times10^9 & 0 & 0 & -1.12\times10^9 & 0 & 0 \\ 0 & 3.19\times10^6 & 4.7\times10^6 & 0 & -3.19\times10^6 & 4.7\times10^6 \\ 0 & 4.7\times10^6 & 9.5\times10^6 & 0 & -4.75\times10^6 & 4.7\times10^6 \\ -1.12\times10^9 & 0 & 0 & 1.121\times10^9 & 0 & 2.39\times10^6 \\ 0 & -3.19\times10^6 & 4.7\times10^6 & 0 & 1.121\times10^9 & 4.75\times10^6 \\ 0 & 4.7\times10^6 & 4.75\times10^6 & 2.39\times10^6 & 4.75\times10^6 & 19\times10^6 \end{bmatrix} \times \begin{Bmatrix} u_1 \\ v_1 \\ \theta_1 \\ u_2 \\ v_2 \\ \theta_2 \end{Bmatrix}$$

$$+ \begin{bmatrix} 0 & 0 & 0 \\ 0 & 0 & 0 \\ 0 & 0 & 0 \\ -770\times10^3 & 0 & 2.3\times10^6 \\ 0 & -1.12\times10^{10} & 0 \\ -2.39\times10^6 & 0 & 4.75\times10^6 \end{bmatrix} \times \begin{bmatrix} 0 \\ 0 \\ 0 \end{bmatrix} = \begin{Bmatrix} 0 \\ -60,000 \\ 0 \\ 0 \\ 0 \\ 0 \end{Bmatrix}$$

$$\begin{bmatrix} 1.12\times10^9 & 0 & 0 & -1.12\times10^9 & 0 & 0 \\ 0 & 3.19\times10^6 & 4.7\times10^6 & 0 & -3.19\times10^6 & 4.7\times10^6 \\ 0 & 4.7\times10^6 & 9.5\times10^6 & 0 & -4.75\times10^6 & 4.7\times10^6 \\ -1.12\times10^9 & 0 & 0 & 1.121\times10^9 & 0 & 2.39\times10^6 \\ 0 & -3.19\times10^6 & 4.7\times10^6 & 0 & 1.121\times10^9 & 4.75\times10^6 \\ 0 & 4.7\times10^6 & 4.75\times10^6 & 2.39\times10^6 & 4.75\times10^6 & 19\times10^6 \end{bmatrix}$$

$$\times \begin{Bmatrix} u_1 \\ v_1 \\ \theta_1 \\ u_2 \\ v_2 \\ \theta_2 \end{Bmatrix} = \begin{Bmatrix} 0 \\ -60,000 \\ 0 \\ 0 \\ 0 \\ 0 \end{Bmatrix}$$

$$\{\Delta_u\} = \begin{Bmatrix} u_1 \\ v_1 \\ \theta_1 \\ u_2 \\ v_2 \\ \theta_2 \end{Bmatrix} = \begin{Bmatrix} -0.0916 \\ -0.1756 \\ 0.0674 \\ -0.0916 \\ -0.0009 \\ 0.0383 \end{Bmatrix}$$

$$[k_{21}]\{\Delta_u\} + [k_{22}]\{\Delta_k\} = \{F_u\}$$

$$\begin{bmatrix} 0 & 0 & 0 & -7.9\times10^{10} & 0 & 2.39\times10^6 \\ 0 & 0 & 0 & 0 & -1.12\times10^9 & 0 \\ 0 & 0 & 0 & -2.39\times10^6 & 0 & 4.75\times10^6 \end{bmatrix} \times \begin{Bmatrix} u_1 \\ v_1 \\ \theta_1 \\ u_2 \\ v_2 \\ \theta_2 \end{Bmatrix}$$

$$+ \begin{bmatrix} 770\times10^3 & 0 & 2.3\times10^6 \\ 0 & 120\times10^6 & 0 \\ 2.39\times10^{11} & 0 & 9.56\times10^6 \end{bmatrix} \times \begin{Bmatrix} 0 \\ 0 \\ 0 \end{Bmatrix} = \begin{Bmatrix} F_{3x} \\ F_{3y} \\ M_3 \end{Bmatrix}$$

$$\{F_u\} = \begin{Bmatrix} F_{3x} \\ F_{3y} \\ M_3 \end{Bmatrix} = \begin{Bmatrix} 7.58\times10^9 \\ -1.008\times10^6 \\ -36.999\times10^3 \end{Bmatrix}$$

Questions

1. Develop the stiffness equation.
2. Give step-by-step method of analysis of indeterminate structure by stiffness method.
3. Analyse the frame shown in Fig. 3.20 by stiffness matrix method.

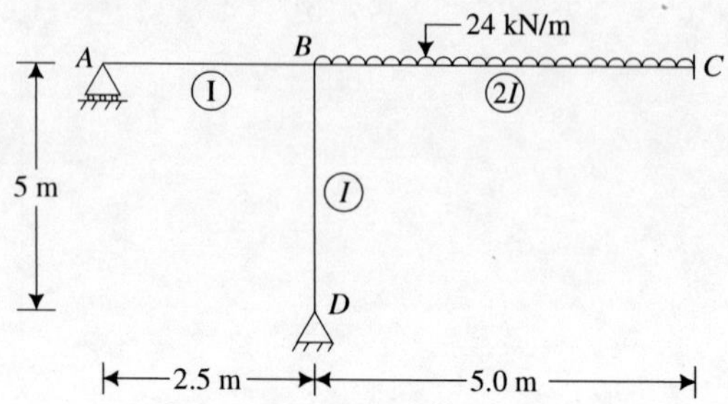

Fig. 3.20 Question No. 3.

Ans. $M_{AB} = 0$, $M_{BA} = 17.65$ kN-m, $M_{BC} = -26.47$ kN-m, $M_{CB} = 61.76$ kN-m, $M_{BD} = 8.82$ kN-m, $M_{DB} = 0$

4. Analyse the frame shown in Fig. 3.21 by stiffness matrix method if support *D* settles by 100/*EI* units horizontally towards right and 50/*EI* units vertically downward. Assume same flexural rigidity for all members.

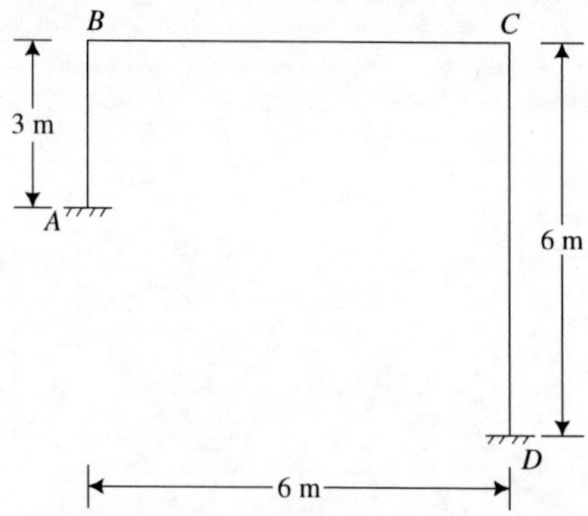

Fig 3.21 Question No. 4.

5. Assemble global stiffness matrix for the plane truss shown in Fig. 3.22. Take $\frac{EA}{L} = 25$ kN-m all member.

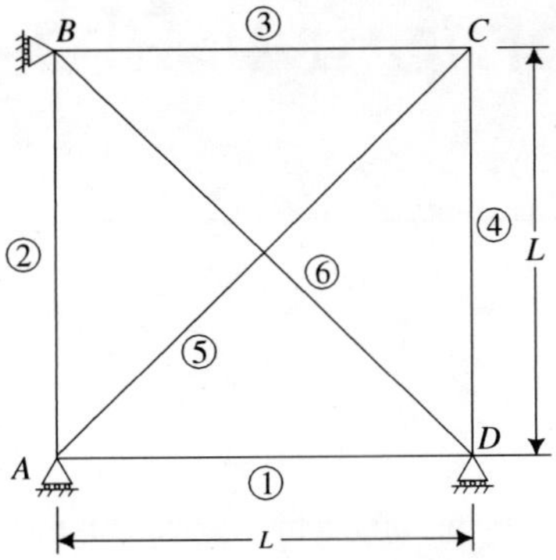

Fig. 3.22 Question No 5.

Ans:

$$25\begin{bmatrix} 1.5 & 0.5 & 0 & 0 & -0.5 & -0.5 & -1.0 & 0.5 \\ & & 1.5 & 0 & -1.0 & -0.5 & -0.5 & 00 \\ & & 1.5 & -0.5 & -1.0 & 0 & -0.5 & 0.5 \\ & & & 1.5 & 0 & 0 & 0.5 & -0.5 \\ & & \text{sym.} & & 1.5 & 0.5 & 0 & 0 \\ & & & & & 1.5 & 0 & 0 \\ & & & & & & 1.5 & -0.5 \\ & & & & & & & 1.5 \end{bmatrix}$$

6. Analyse the continuous beam shown in Fig. 3.23 by stiffness matrix-element approach. *EI* is constant throughout.

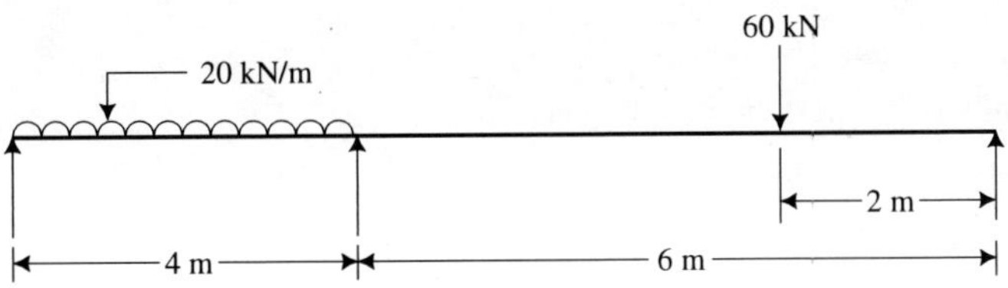

Fig. 3.23 Question No. 6.

4 Computer Programming Preliminaries

To make the methods clear, few problems have been solved by hand calculations in the last two chapters. The matrix methods are not meant for hand calculations. The element approaches of matrix methods are ideally suited for developing general-purpose computer programs. In this chapter the programming preliminaries for stiffness matrix method are discussed so that one can develop his/her own computer program or if he/she is user of a developed program, can understand how to use the problem. On the same line the computer program for flexibility matrix method may be developed.

4.1 INPUT

Various input information required may be grouped into

(a) Geometric data
(b) Load data
(c) Material properties
(d) Boundary conditions

4.1.1 Geometric Data

The computer is to be informed about the geometry of the structure to be analysed clearly by giving suitable data. This data may be grouped into

(i) General information
(ii) Coordinates of nodes
(iii) Nodal connectivity details
(iv) Details of properties of cross sections.

(i) General Information

It includes giving input regarding

NE – total number of elements

NN – total number of nodes

NDF – number of degrees of freedom at each node

(ii) Coordinates of Nodes

Coordinates of each node are either to be supplied or generated. Coordinates may be stored in the form

CORD (i, 1) – x coordinate of node i

CORD (i, 2) – y coordinate of node i

CORD (i, 3) – z coordinate of node i

If the coordinates are to be generated, then instead of reading coordinates, statement is given to enter a user written subprogram, in which once a few dimensions are supplied, coordinates of all nodes are calculated.

(iii) Nodal Connectivity Details

It involves in giving information about the end nodes of an element. For each member read NENDI and NENDJ. While reading this information, the material member may also be read, so that a structure with different materials also may be analysed.

(iv) Cross-Sectional Properties

This information about each element is required so as to find moment of inertia about required axes and the area of cross section. The values may be directly supplied or calculated.

Figures 4.1 and 4.2 show two typical problems and input information to be given.

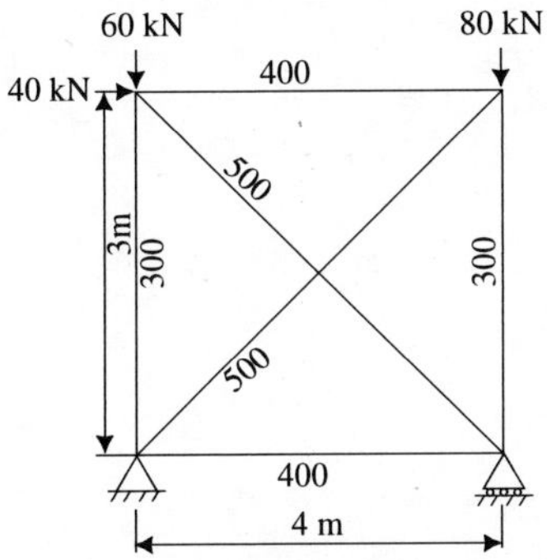

(a) Given problem

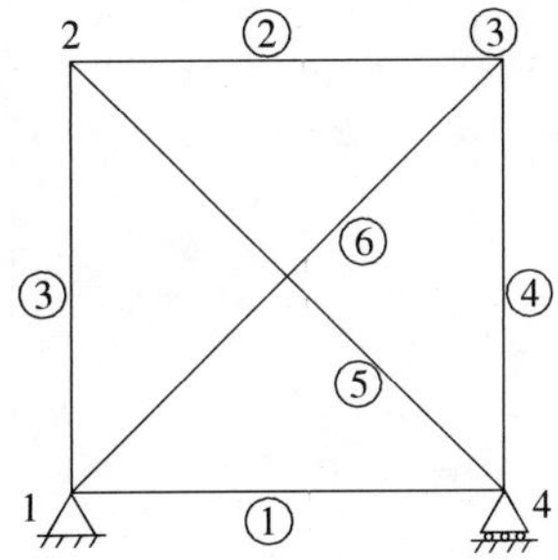

(b) Member and node numbers

Fig. 4.1

Input for problem 1

NE = 6, NN = 4, NDF = 2

CORD (1 ,1) = 0.0 CORD (1, 2) = 0.0

CORD (2, 1) = 0.0 CORD (2, 2) = 3.0 m

CORD (3, 1) = 4.0 m CORD (3, 2) = 3.0 m

CORD (4, 1) = 4.0 m CORD (4, 2) = 0.0

Nodal connectivity details and material number

Element No.	NENDI	NENDJ	Material Number
1	1	4	1
2	2	3	1
3	1	2	1
4	4	3	1
5	4	2	1
6	1	3	1

Cross-sectional property:

Element no.	Area of cross section (in m^2)
1	400×10^{-6}
2	400×10^{-6}
3	300×10^{-6}
4	300×10^{-6}
5	500×10^{-6}
6	500×10^{-6}

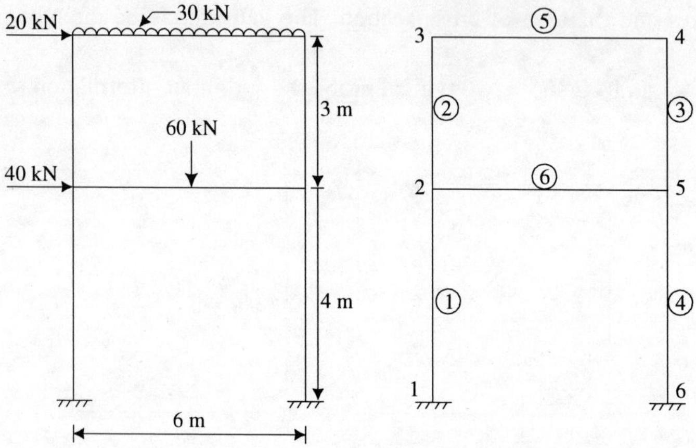

(a) Given problem   (a) Member and node numbers

Fig. 4.2

Input for problem 2

NE = 6   NN = 6   NDF = 3

CORD (1, 1) = 0.0   CORD (1, 2) = 0.0

CORD (2, 1) = 0.0   CORD (2, 2) = 4.0 m

CORD (3, 1) = 0.0   CORD (3, 2) = 7.0 m

CORD (4, 1) = 6.0 m   CORD (4, 2) = 7.0 m

CORD (5, 1) = 6.0 m   CORD (5, 2) = 4.0 m

CORD (6, 1) = 6.0 m   CORD (6, 2) = 0.0

Nodal connectivity details and material number

Element no.	NENDI	NENDJ	Material No.
1	1	2	1
2	2	3	1
3	5	4	1
4	6	5	1
5	3	4	1
6	2	5	1

Cross-sectional details: If columns are of size 300 mm × 300 mm and beams are of size 300 mm × 450 mm

Element no.	Area of C.S. (in m^2)	$I = I_{22}$ (in m^4)
1	0.09	2.25×10^{-6}
2	0.09	2.25×10^{-6}
3	0.09	2.25×10^{-6}
4	0.09	2.25×10^{-6}
5	0.135	2.278×10^{-6}
6	0.135	2.278×10^{-6}

4.1.2 Load Data

Loads may be acting on the nodes or on the member.

For loads acting on nodes, the following input is to be given:

- Number of such loads.
- For each load case give joint number, magnitude of load, coordinate direction.

For loads acting on member, the information to be given is:

- Number of elements subject to such loads.
- For each element subject to such loads give load type, magnitude of load, distance from ENDI if required.

For example, for *problem 1,* load data information

Number of nodes subject to loads on node = 2

Load case	Node no.	Magnitude	Coordinate direction
1	2	–60	2
2	2	40	1
3	3	–80	2

For loads acting on member – No such case. Skip asking information.

For problem 2

For loads acting on nodes.

Number of such loads 2.

Load case	Node no.	Magnitude	Direction
1	3	20	1
2	2	40	1

For loads acting on the member:

Number of elements subject to such loads 2

Elements subject to load	Type	Magnitude	Distance from ENDI
5	2	–30	–
6	1	–60	3.0

[*Note:* Type of load is numbered as per Table A.1]

4.1.3 Material Properties

First read total number of materials. Then for each material read the necessary material properties.

For problem 1,

Number of materials used = 1

Material no.	Young's modulus	Poisson's ratio
1	200×10^6	0.3

For problem 2,

Number of materials used 1

Material No.	Young's modulus	Poisson's ratio
1	15×10^6	0.15

4.1.4 Boundary Conditions

First read number of nodes subject to boundary conditions. Then for each such node, read the boundary condition. If there is no boundary conditions in a particular coordinate direction, one of the methods of identifying it is to put a large number. While imposing boundary conditions, if this large number is encountered skip imposing boundary condition in that coordinate direction.

For problem 1.

No. of nodes subject to boundary conditions = 2.

Node	BOUND (I, 1)	BOUND (I, 2)
1	0.0	0.0
2	1×10^{12}	0.0

For problem 2,

No. of nodes subject to boundary conditions = 2

Node	BOUND (I, 1)	BOUND (I, 2)	BOUND (I, 3)
1	0.0	0.0	0.0
6	0.0	0.0	0.0

4.2 PROCESSING

It involves the following steps:

1. Assemble global stiffness matrix
2. Assemble load vector
3. Impose boundary conditions
4. Solve simultaneous equations
5. Find member end forces.

4.2.1 Assembling Global Stiffness Matrix

Steps involved are:

1. Find size of global stiffness matrix (NSIZE = NNODE × NDF). Initialize NSIZE × NSIZE global stiffness matrix $[k]$.
2. Enter element loop.
   (a) Initialize element stiffness matrix
   (b) Generate element stiffness matrix [k□]
   (c) Generate rotation matrix $[R]$
   (d) Then $[k_m] = [R]^T [k_m] [R]$
   (e) Knowing NENDI and NENDJ of element, identify row and column of each term in element stiffness matrix in the appropriate row and column of global stiffness matrix. Add that term to already existing value in that position.

At the end of element loop, unassembled global stiffness matrix is available.

Note: The steps *b*, *c* and *d* may be replaced by giving element stiffness matrix in global form itself.

4.2.2 Assembling Load Vector

(a) First initialize global load vector $\{F_c\}$

(b) For loads acting on nodes enter a loop to
- identify the global coordinate direction
- add that load to existing load in $\{F_c\}$

(c) For loads acting on elements enter a loop. For each loaded element,
- Initialize $\{F_{mF}\}$ and $\{F_{me}\}$vector
- Knowing the type of load, its magnitude and point of application enter the corresponding segment calculating $\{F_{mF}\}$. Store this vector.
- Find $\{F_{me}\}$ which is equal to $-\{F_{mF}\}$.
- Add each term of $\{F_{me}\}$ to appropriate row of existing values of $[F_c\}$.
- At the end of the loop $\{F_c\}$is available.

4.2.3 Imposing Boundary Conditions

At this stage the equation of $[k]\ \{\Delta\} = \{F_c\}$ is available.

Number of nodes subject to boundary conditions is also known. For each boundary condition, find equation number. If the value of boundary condition is very large specified value, it is identified that there is no boundary condition. If the boundary condition exists for that coordinate direction, impose it by suitably altering values in stiffness matrix and the load vector. At the end of boundary condition loop, appropriate equation is available.

4.2.4 Solving Simultaneous Equations

For solving simultaneous equations, standard packages may be used which use either Gauss elimination technique or cholesky secomposition technique. Care should be taken to see that the form of equation developed matches with that of standard package.

4.2.5 Finding Member End Forces

The solution of simultaneous equations give displacement vector in global system. To find member end forces in the local system

- Enter an element loop
- Initialize $[F_m]$
- Develop $[R]$
- Knowing end numbers of the element pick up $\{\Delta_m\}$
- Find $\{\Delta'_m\} = [R]\ \{\Delta_m\}$
- Find or pickup already stored $[k'_m]$
- Then $\{F_m\} = [k'_m]\ \{\Delta'_m\}$

At the end of the loop member end forces of all elements are available.

Figure 4.3 shows the flowchart.

4.3 OUTPUT

The member end forces are printed out along with element member and coordinate direction. Usually before getting final output title of the problem and input information is also printed out.

4.4 TECHNIQUES OF SAVING COMPUTER MEMORY REQUIREMENTS

Memory requirement for storing stiffness matrix of NSIZE × NSIZE is very large. Hence, a user may end up with shortage of computer active memory while handling large problems. The following techniques have been developed to overcome this problem:

(i) Use of symmetry and banded nature of stiffness matrix

(ii) Partitioning of the matrix

(iii) Skyline storage

4.4.1 Use of Symmetry and Banded Nature

It is obvious from Maxwell's reciprocal theorem that stiffness matrix is a symmetric matrix. Hence, only upper or lower triangular matrix may be developed. Apart from this the matrix has banded nature, i.e., the non-zero elements are concentrated near the diagonal of the matrix as shown in Fig. 4.4. Hence, instead of storing NSIZE × NSIZE stiffness matrix only NSIZE × B size matrix may be stored and the computations managed. The semi-band width B may be found as given below:

$$B = [\text{Max of } (|\text{NENDI}-\text{NENDJ}|) + 1]\ \text{NDF}$$

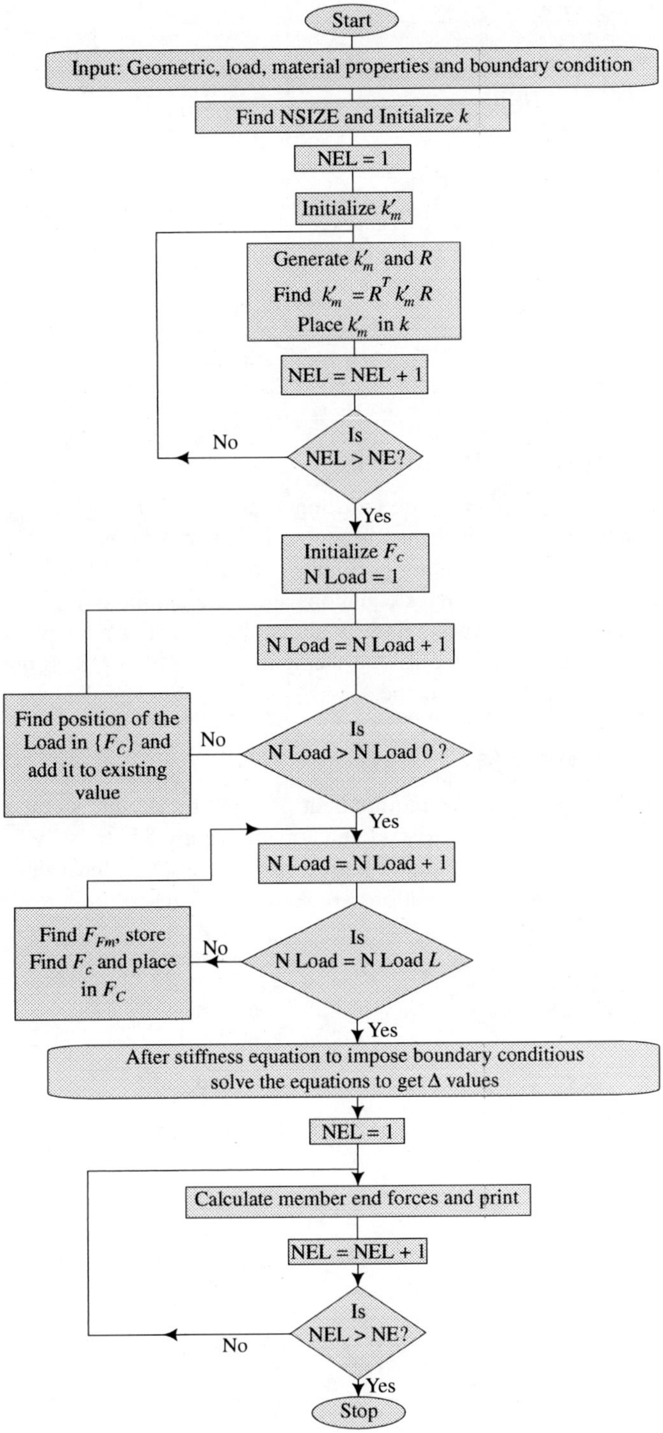

Fig. 4.3 Flowchart.

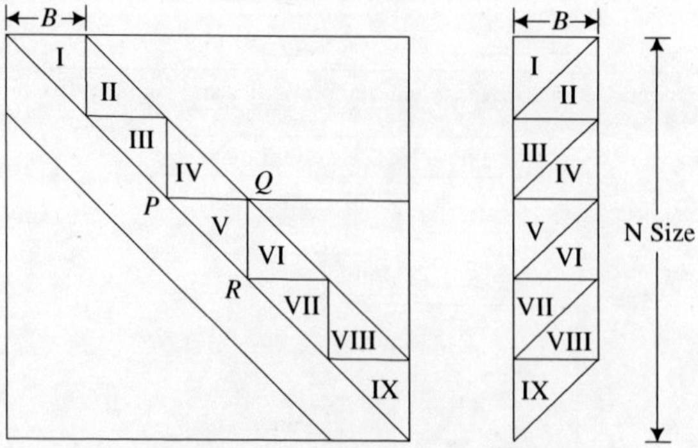

Fig. 4.4 Banded nature of stiffness matrix.

4.4.2 Partitioning of the Matrix

The banded stiffness matrix may be partitioned as shown in Fig. 4.3. Then only a few of the triangular submatrices may be stored in the active memory of computer at any given time, while the remaining portions are dumped on hard disk. It is possible to manage solution of simultaneous equations since the elimination performed using one row affects only a few rows below or above that row. For example, elimination operations in row PQ modifies only the elements in triangle PQR. Frontal solution technique is developed on this scheme.

4.4.3 Skyline Storage

Further saving in memory is the requirement possible by skyline storage technique. In this system if zero values exist at the top of bandwidth those values are not stored. The line marked in Fig. 4.5 is known as skyline. Only the values below the skyline are stored and matrix operations are managed to get solution to simultaneous equations.

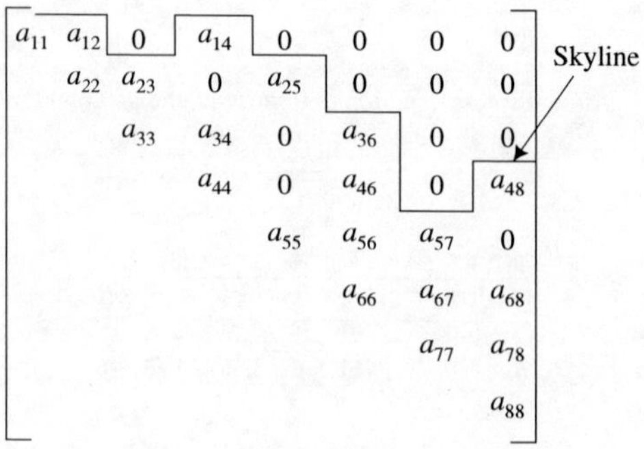

Fig. 4.5 Skyline storage technique.

4.5 SOFTWARE DEVELOPERS TO NOTE

Software developers of analysis package should note the following points.

1. Use right-hand rule for selecting x, y, and z coordinates. The rule says, if thumb, index finger and middle finger are stretched so as to show three mutually perpendicular directions, thumb shows x-direction, index finger shows y-direction, and middle finger shows z-direction.
2. Forces acting in the positive direction of coordinates are taken as positive forces.
3. Moments are referred about an axis, not in the direction of an axis. If thumb of right hand is stretched and held towards positive direction of a coordinate, the closed fingers indicate positive sense of moment about the axis. Thus, in a beam, if x-axis towards right, y-axis is upward, anticlockwise moment is positive M_Z (Refer Fig. 4.6).

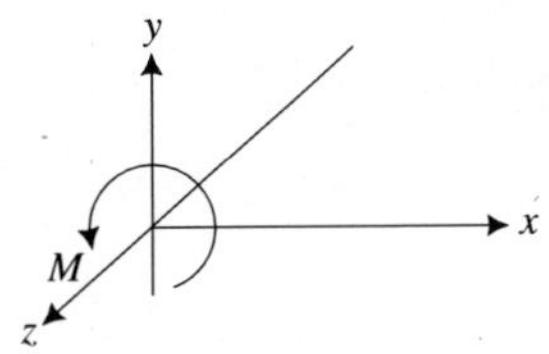

Fig. 4.6 Positive direction of moment, *M*.

4. Use consistency of units, i.e., avoid using metre for some quantities and mm units for some other. Similarly, avoid using kN units for some quantities and Newton's for some quantities.
5. Test the program developed with a number of simple test problems for which results are available by classical theories.

4.6 SOFTWARE USERS TO NOTE

The software users should note the following:

1. A program developed with right-hand rule for coordinate hold good for any orientation of coordinates.
2. If the program has built-in self weight calculation, the coordinate orientation should match with that used by the program developers.
3. However, it is a better practice to check the coordinates direction used, since a program developer may or may not follow right-hand rule for coordinates.
4. Check whether self weight has been built in the program and is automatically accounted.
5. Always have own test problems and check whether you are using the software correctly.
6. An analyst should have feel of the structure with a set of drastic assumptions he should visualize deflected surface and identify zones of tension and zones of compression. After getting computer output for a problem, he should check whether the result shows the expected trend.

It should be noted that well developed and used computer program yields accurate results. If mistakes are made in understanding positive senses of vectors (displacements, loads, forces, moments, etc.) the results obtained will be absurd. The user of software should develop the capacity to feel the structure and easily identify if the results are absurd.

ASSIGNMENT

Develop computer programs for the analysis of the following structures with NNODES and NE elements:

1. Two-dimensional trusses
2. Space trusses
3. Continuous beams
4. Rigid jointed plane frames
5. Rigid jointed space frames
6. Grids

Appendix: Matrix Algebra Review

PRELIMINARIES

A matrix is a way of organizing information.

It is a rectangular array of elements arranged in rows and columns. For example, the following matrix A has m rows and n columns.

$$A = \begin{bmatrix} a_{11} & a_{12} & a_{13} & . & . & a_{1n} \\ a_{21} & a_{22} & a_{23} & . & . & a_{2n} \\ a_{31} & a_{32} & a_{33} & . & . & a_{3n} \\ . & . & . & . & . & . \\ . & . & . & . & . & . \\ a_{m1} & a_{m2} & a_{m3} & . & . & a_{mn} \end{bmatrix}$$

All elements can be identified by a typical element a_{ij}, where $i = 1, 2, \ldots, m$ denotes rows and $j = 1, 2, \ldots, n$ denotes columns.

A matrix is of order (or dimension) m by n (also denoted as $(m \times n)$).

A matrix that has a single column is called a column vector.

A matrix that has a single row is called a row vector.

MATRIX TRANSPOSE

The transpose of a matrix or vector is formed by interchanging the rows and the columns. A matrix of order $(m \times n)$ becomes of order $(n \times m)$ when transposed.

For example, if A (2×3) matrix is defined by

$$A = \begin{bmatrix} a_{11} & a_{12} & a_{13} \\ a_{21} & a_{22} & a_{23} \end{bmatrix}$$

Then the transpose of A, denoted by A', is now (3×2)

$$A' = \begin{bmatrix} a_{11} & a_{21} \\ a_{12} & a_{22} \\ a_{13} & a_{23} \end{bmatrix}$$

SYMMETRIC MATRIX

When $A' = A$, the matrix is called **symmetric**. That is, a symmetric matrix is a square matrix, in that it has the same number of rows as it has columns, and the off-diagonal elements are symmetric

(i.e., $a_{ij} = a_{ji}$ for all i and j)

For example,

$$A = \begin{bmatrix} 4 & 5 & -3 \\ 5 & 7 & 2 \\ -3 & 2 & 10 \end{bmatrix}$$

A special case is the **identity matrix**, which has 1's on the diagonal positions and 0's on the off-diagonal positions.

$$I = \begin{bmatrix} 1 & 0 & 0 \\ 0 & 1 & 0 \\ 0 & 0 & 1 \end{bmatrix}$$

MATRIX ADDITION AND SUBTRACTION

Matrices can be added and subtracted as long as they are of the same dimension. The addition of matrix A and matrix B is the addition of the corresponding elements of A and B. So, $C = A + B$ implies that $c_{ij} = a_{ij} + b_{ij}$ for all i and j.

For example,

$$A = \begin{bmatrix} 2 & -3 \\ 6 & 10 \end{bmatrix}$$

$$B = \begin{bmatrix} 0 & 6 \\ 5 & -8 \end{bmatrix}$$

Then, $C = A + B = \begin{bmatrix} 2 & -3 \\ 6 & 10 \end{bmatrix} + \begin{bmatrix} 0 & 6 \\ 5 & -8 \end{bmatrix}$

$$C = \begin{bmatrix} 2 & 3 \\ 11 & 2 \end{bmatrix}$$

MATRIX MULTIPLICATION

In case of multiplying two matrices, such as $C = AB$, it must be a case that:

The number of columns of A = The number of rows of B.

So, if A is of dimension $(m \times p)$ and B of dimension $(p \times n)$, then the product, C, will be of order $(m \times n)$ whose ij^{th} element is defined as $C_{ij} = \sum_{k=1}^{p} A_{ik} B_{kj}$.

In other words, the ij^{th} element of the product matrix is found by multiplying the elements of the ith row of A, the first matrix, by the corresponding elements of the jth column of B, the second matrix and summing the resulting product. For this to hold good, the number of columns in the first matrix must be equal the number of rows in the second.

For example,

$$F = AD = \begin{bmatrix} 6 & 8 \\ -2 & 4 \end{bmatrix} * \begin{bmatrix} 3 & -8 & 1 \\ 9 & 2 & 5 \end{bmatrix}$$

$$= \begin{bmatrix} 6*3+8*9 & 6*(-8)+8*2 & 6*1+8*5 \\ (-2)*3+4*9 & (-2)*(-8)+4*2 & (-2)*1+4*5 \end{bmatrix}$$

$$= \begin{bmatrix} 90 & -32 & 46 \\ 30 & 24 & 18 \end{bmatrix}$$

Determinant of a Square Matrix

The determinant of a square matrix A, is denoted by $|A|$

(i) For a single element matrix, $|A| = a_{11}$

(ii) In the case of 2 × 2 matrix

$$A = \begin{bmatrix} a_{11} & a_{12} \\ a_{21} & a_{22} \end{bmatrix}$$

The determinant is defined to be the difference of two terms as follows:

$$|A| = a_{11}\, a_{22} - a_{12}\, a_{21}$$

which is obtained by multiplying the two elements in the principal diagonal of A and then subtracting the product of the two off-diagonal elements.

(iii) In the case of 3 × 3 matrix

$$A = \begin{bmatrix} a_{11} & a_{12} & a_{13} \\ a_{21} & a_{22} & a_{23} \\ a_{31} & a_{32} & a_{33} \end{bmatrix}$$

$$|A| = a_{11} \begin{vmatrix} a_{22} & a_{23} \\ a_{32} & a_{33} \end{vmatrix} - a_{12} \begin{vmatrix} a_{21} & a_{23} \\ a_{31} & a_{33} \end{vmatrix} + a_{13} \begin{vmatrix} a_{21} & a_{22} \\ a_{31} & a_{32} \end{vmatrix}$$

Inverse of a Matrix

The inverse of a matrix A is obtained from the following formula

$$A^{-1} = \frac{C'}{|A|} \text{ where } C' = [c_{ij}] \text{ and } c_{ij} = (-1)^{i+j} |A_{ij}|$$

For example, let

$$A = \begin{bmatrix} 1 & 2 \\ 3 & 4 \end{bmatrix}$$

$$|A| = -2$$

And the cofactors are $c_{11} = 4$; $c_{22} = 1$; $c_{12} = -3$ and $c_{21} = -2$

$$A^{-1} = \frac{1}{-2} \begin{bmatrix} 4 & -3 \\ -2 & 1 \end{bmatrix} = \begin{bmatrix} -2 & 1 \\ 1.5 & -0.5 \end{bmatrix}$$

Index